Rebuilding the Earth

"This is an important and timely book. Mark's message is clear; it is time to rebuild the Ark. The blueprint for the construction of the Ark is available; Mark provides much needed positive examples of how people have chosen a more sustainable path. Crucially this book is about hope—hope for the planet, hope for nature and hope for humanity."
—Dr Gary Mantle MBE, *Chief Executive of Wiltshire Wildlife Trust*

"With the overall pessimism regarding the future of our planet, this book brings some hope that we can still put the boat back to rights, enlightening our paths and actions through the global ecosystem's metabolism and capacity to rejuvenate. Despite the irresponsible leadership being demonstrated across the globe with regard to the destruction of our planet's resources and ecosystems, the *sine qua non* condition for life on Earth, as well as towards our current and future wellbeing, Mark attempts, with persistence and perseverance based on good practice, to demonstrate how we can still recover and reactivate the vital functionalities of system Earth—a highly commendable message of hope. In my long career promoting sustainable consumption and production patterns, I have encountered far more frustration than effective transformative change, but I am willing to share Mark's optimism. Let us give it a try, believe and act with an attitude of 'Yes, We Can' for the future of Earth, people and humanity."
—Arab Hoballah, *Former Chief of Sustainable Consumption and Production/SCP, United Nations Environment Programme (UNEP)*

"The great thing about Mark Everard's writing is that he combines the philosophical with the practical, making you think more deeply about issues vital to the future wellbeing of the planet—but then he provides potential solutions, and is even optimistic that we can actually deliver some of them and save the environment, on which we are all completely dependent for our own existence, from the havoc we humans are wreaking across the entire globe."
—Paul Knight, Director, *Salmon and Trout Conservation, UK*

"Here's the thing: we absolutely know how to restore all the Earth's critical ecosystems. A combination of ongoing inspired research, by experienced scientists such as Dr Everard, plus a reservoir of 'recovered wisdom' from people with deep knowledge of the places in which they live—today's 'citizen scientists', if you like—makes this regenerative agenda practical, scaleable and transformational right now—after decades of degeneration and devastation.

But we have to think about these challenges so much more holistically. Astonishingly, today's climate experts populate one big silo, and biodiversity experts another! The notion of 'co-benefits' barely begins to do justice to the emerging revolution in both thinking and practice stimulated by the realisation that we face simultaneous emergencies on both fronts. Which means quite simply that natural climate solutions must become the watchword of the next decade."
—Jonathon Porritt, Founder Director, *Forum for the Future*

"If a common good comes from our daily twitter onslaught of sobering environmental news, it is the realisation that our connectedness can potentially lead to sea changes in the approach to pushing back against the damage. From out of this news bombardment we need to chart reasons for hope, and examples and pathways through the maze of problems that may appear to be intractable. This book offers a well-balanced picture of the challenges, and the reasons to be hopeful—that we can do more than we have yet been led to believe.

This book is a refreshing take on the foremost challenge of humanity. It is a call to embrace opportunities and is well grounded in the delivery of examples—from all nations and continents undergoing different degrees of development—as to regenerative approaches at scales appropriate to the landscapes in question.

Calling upon our inner and innate need for survival, the author sets a positive and inclusive tone with a sense that we can learn, contribute and give back to our planet. Dr Everard shows us that solutions are there, and that we have the tools and the intelligence to revive and restore our natural systems through our own ingenuity and experience. Rebuilding and reconnecting to Earth is conveyed as nothing less than a mission; one that is far from impossible."

—Stuart Orr, Freshwater Practice Leader, *WWF International*

"A well-written book which communicates powerful messages, based largely upon the author's own experiences and a lifetime of professional expertise. This work doesn't simply identify and describe the problems with Earth's ecosystems and the services they supply, but also describes realistic solutions based on the premise of reviving and rescuing the natural world in a stepwise manner. Accessible yet erudite, this is a must read for anyone interested in learning how we can begin to reverse the damage we have done to the planet's ecosystems."

—Dr Paul Johnston, Principal Scientist, *Greenpeace International Science Laboratory*

"Sustainable development cannot be achieved simply by 'lightening our load' on a seriously degraded and declining natural world. 'Rebuilding the Earth' highlights both humanity's complete interdependence with the natural world, and that any serious engagement with sustainable development must entail rebuilding the foundering structure and functioning of nature as the irreplaceable bedrock of future human security and progress. The book is above all optimistic about the daunting challenges facing global society, its comprehensive range of case studies showing that we not only can but already are making regenerative change in fragmented ways across the world."

—Professor James W. S. Longhurst, *Professor of Environmental Science and Assistant Vice Chancellor for Environment and Sustainability, University of the West of England (UWE, Bristol)*

"Regenerating natural spaces and systems spoiled by human blindness is a critical element of any future that we could call desirable. Dr Everard's book makes a relevant and inspiring contribution to put the topic much higher in all our agendas. It paves as well the way to reconsider our interdependencies with what we call nature, actually part of ourselves. Hopefully his call will be heard and help to erase the split between society and nature, urban and rural, our wellbeing and the healthy biosphere we all need to keep alive."

—Carlos Alvarez Pereira, *Member of the Executive Committee of the Club of Rome*

Mark Everard

Rebuilding the Earth

Regenerating our planet's life support
systems for a sustainable future

palgrave
macmillan

Mark Everard
University of the West of England
Bristol, UK

ISBN 978-3-030-33023-1   ISBN 978-3-030-33024-8   (eBook)
https://doi.org/10.1007/978-3-030-33024-8

Cover illustration: © Borchee/gettyimages

This Palgrave Macmillan imprint is published by the registered company Springer Nature Switzerland AG.
The registered company address is: Gewerbestrasse 11, 6330 Cham, Switzerland

Acknowledgements

The author is grateful to the RICS Research Trust, the Lloyds Register Foundation, the John Pontin Trust and the University of the West of England (UWE Bristol) for funding elements of travel undertaken in background research. The bulk of research time resulted from private scholarship, and I am grateful for the forbearance and support from my family, Jackie and Daisy, including valuable technical critique from Jackie.

Research trips in India included study of socio-ecological regeneration schemes operated by the NGOs Tarun Bharat Sangh (TBS), WaterHarvest (formerly known as Wells for India) and GVNML in Rajasthan. Trips also included multiple visits to the WaterHarvest office in the UK, the Pitchandikulam Bio Resources Centre in Tamil Nadu, schemes in the Middle Himalayas associated with research links with Kumaun University, investigation of the impact of protected tribal rights on ecosystem-livelihood interdependencies in Arunachal Pradesh, and opportunist visits to Wetlands International—South Asia offices in New Delhi.

In India, my thanks go to my friends and colleagues at TBS (Rajendra Singh, Kanhaiya Lal, Gopal Singh, Maulik Sisodia and Suresh Raikwar), Rudhmal Mena (headman of Harmeerpur Village), Om Prakash Sharma (Country manager, WaterHarvest), Lakshman Singh and Jagveer Singh (GVNML), Rakesh Vaish for transport, translation and explanation of local contexts as well as Gaurav Kataria, Smita Kumar and all at AE Travel Pvt Ltd. Down in Tamil Nadu, thanks to Joss Brooks at the Pitchandukulam Bio Resources Centre, Auroville.

RICS Research Trust support also enabled me to present some of the findings of initial research on regenerative Indian ecosystem-livelihood interactions at the 5th international EcoSummit conference (Montpellier, France, August–September 2016). Additional major international conferences at which learning from this work has been presented include the Society of Wetlands Sciences (SWS) 11th Annual European Chapter meeting (Potsdam, Germany, May 2016) and the International Ground Water Conference 2017 (IGWC2017, New Delhi, December 2017).

Beyond the fieldwork and the institutions, there are too many people to thank individually. But special thanks anyhow to Rob McInnes for constant intellectual exchange in the work we do together around the world. Also to Dr John Colvin and Professor Jim Longhurst for great input of ideas on the journey.

Finally, thanks to Rob McInnes, Daisy Everard and Jagveer Singh for permission to use relevant photographs, as credited in Figure legends.

Contents

Acronyms

ACWADAM	Advanced Centre for Water Resources Development and Management
AFR100	African Forest Landscape Restoration Initiative
AR	Artificial Aquifer Recharge
ARLI	African Resilient Landscapes Initiative
ASR	Aquifer Storage and Recovery
B2C2	Bhutan Biological Conservation Complex
BiTC	Business in the Community
CIF	Climate Investment Funds
CRP	US Conservation Reserve Program
eNGO	Environment NGO
FAP	World Bank Group Forest Action Plan
GPS	Global Positioning System
GVNML	Gram Vikas Navyuvak Mandal, Laporiya (Indian NGO)
ICW	Integrated constructed wetland
IDA	International Development Association
IMAWESA	Improved Management of Agricultural Water in Eastern and Southern Africa
IWSN	International Water Security Network
JSA	Jalyukt Shivar Abhiyan
LDBA	Biodiversity and Land Degradation
MAR	Managed Aquifer Recharge
MDB	Multilateral development bank

MJSA	*Mukhya Mantri Jal Swavlamban Abhiyan*, the 'Chief Minister's water self-sufficiency mission' of the Government of Rajasthan
MRSAC	Maharashtra Remote Sensing Application Centre
Neoliberal	A modified form of liberalism favouring free-market capitalism
NEPAD	New Partnership for Africa's Development
NFM	Natural flood management
NGO	Non-governmental organisation
PES	Payment for ecosystem services
PROFOR	The Program on Forests
REDD+	The United Nations Collaborative Programme on Reducing Emissions from Deforestation and Forest Degradation in Developing Countries
SCaMP	Sustainable Catchment Management Programme
SES	Socio-ecological systems
STEEP	A conceptual framework comprising social, technological, economic, environmental and political elements
SuDS	Sustainable drainage systems
TBS	Tarun Bharat Sangh (Indian NGO)
TCW	The Converging World (NGO)
TDEF	Tropical dry evergreen forest
UNCCD	UN Convention on Combating Desertification
US EPA	United States Environmental Protection Agency
VDC	Village Development Committee
WBCSD	World Business Council for Sustainable Development
WHS	Water-harvesting structure

List of Figures

List of Tables

List of Boxes

1

Introduction

Human wellbeing and future prospects depend totally upon the supportive capacities of the natural world, with which we co-evolved as integral elements. Yet societal habits are shockingly naïve about this fact. Lack of consideration of our reliance and inadvertent impacts upon supporting ecosystems underlies many of today's pressing sustainability challenges.

1.1 Our Ecological Lives

The bare facts of our dependence on ecosystems are as stark as they are inescapable. Humans are biological entities, co-evolved with and entirely dependent on fundamental exchanges with planetary ecosystems. As adults, we breathe roughly 14 breaths each minutes when at rest exchanging approximately 7 litres every minute,[1] cumulatively exchanging well in excess of one-quarter of a billion litres (the capacity of 100+ Olympic-sized swimming pools) with the atmosphere over a 70-year lifetime; deprived of air, we suffer irreversible brain damage in around four min-

---

[1] https://www.britannica.com/science/human-respiratory-system/The-respiratory-pump-and-its-performance.

© The Author(s) 2020
M. Everard, *Rebuilding the Earth*, https://doi.org/10.1007/978-3-030-33024-8_1

1

utes with death ensuing thereafter. We excrete 2–2.5 litres of water each day as urine with more lost as sweat, faeces and moist exhalations, requiring us to replace a minimum of 2.7–3.7 litres daily for optimal health; without fresh water, our cells start to degrade catastrophically after around four days, depending on climate and activity level. Though we may dress it up as gastronomic sophistication, we essentially have to push a diversity of plant, animal and fungal matter down our throats every day to supply sufficient carbohydrate, protein, fat, vitamins and trace minerals to keep our cells in good shape. Denied food, our bodies progressively remobilise tissues, prioritising the needs of the brain and other crucial organs but eventually leading to systemic failure and death after around forty days.

Our broader day-to-day needs are equally ecologically dependent. Servicing our biophysical subsistence needs requires us to harness nature's energy and material flows. But businesses too have a metabolic exchange with energy and material ultimately sourced from natural systems. Large swathes of society accept capitalist business as a favoured model for converting raw resources into useful products to meet their diverse needs and demands. It is regrettable that these dependencies on ecosystems are underappreciated and overlooked across so much of society's diverse activities as, by oversight, we tend inadvertently to degrade this foundational natural capital.

The extent to which we fail to value foundational natural systems, as compared to the things that we build to exploit them, is encoded in generally accepted understandings of the word 'infrastructure' as well as the workings of the market.

1.2   Critical Infrastructure and Market Drivers

UK government is amongst many across the world defining 'critical national infrastructure' in purely man-made terms (see Box 1.1). Substantial economic value is ascribed to the built systems upon which the smooth operation of today's densely clustered human populations substantially depends. Yet, despite its ingenuity, engineered infrastructure is merely a technical means to connect us with natural systems. These include natural flows of energy (much of it today sequestered in fossil reserves), aquatic ecosystems as both water sources and places to dissipate wastes, or to convey food and other resources to us or allow us to access places of aesthetic, economic, spiritual, cultural, educational, recreational or other benefit.

> **Box 1.1 UK Definition of Critical 'National Infrastructure'[2]**
>
> In government terms, 'national infrastructure' comprises facilities, systems, sites, information, people, networks and processes necessary for a country to function and upon which daily life depends. It also includes some functions, sites and organisations which, though not critical to the maintenance of essential services, need protection due to potential public danger (such as civil nuclear and chemical sites).
>
> The UK's *Centre for the Protection of National Infrastructure* defines thirteen national infrastructure sectors: chemicals, civil nuclear communications, defence, emergency services, energy, finance, food, government, health, space, transport and water. Several sectors have defined 'sub-sectors'; emergency services, for example, can be split into police, ambulance, fire services and Coast Guard.

Too often, we omit to reflect that the primary purpose of many technologies is to extend our capacities to access nature's more fundamental infrastructure. Natural systems, albeit many now heavily modified to maximise a subset of desired outputs, are the ultimate source of the food we eat, water put to many consumptive uses as well as a receptor of wastewater, flows of energy, natural regulation of flooding, climate and air quality, and places for recreation, education and other aspects of life fulfilment. Conspicuously lacking from our definition and common appreciation of critical infrastructure are rivers, aquifers, floodplains and wetlands, forests and oceans, green spaces, fertile soils and the atmosphere. All comprise foundational natural infrastructure without which our needs would not merely be compromised, but could not be supported by even the most sophisticated machinery.

---

[2] Centre for the Protection of National Infrastructure [GB]. (2018). *Critical National Infrastructure*. Centre for the Protection of National Infrastructure [GB], HM Government. https://www.cpni.gov.uk/critical-national-infrastructure-0, accessed 28 December 2018.

Nature's 'ark' is indispensable, however much contemporary world views and narrowly framed markets disregard it. Our developed world habit, increasingly pervasive throughout global society, has been to improve our prospects through technically efficient manipulation of ecosystems to maximise the exploitation of a limited subset of natural resources, rather than to recognise and value the ecosystems themselves as the foundations of multiple dimensions of physical, economic and other aspects of wellbeing. The consequent inadvertent serial erosion of natural infrastructure vital for our continued security and opportunity is far from without consequence. Systemically interdependent as we are with the ecosystems providing for our diverse needs, life is not merely impoverished and our resource needs compromised as natural systems decline; in many ways it becomes increasingly impossible.

1.3   Taking a Positive View

The emphasis of this book is upon optimism about the potential to rebuild the natural infrastructure upon which future human opportunity depends. It curates many existing solutions scattered throughout the world where ecosystem enhancement has occurred as a basis for increasing human security and prospects. It seeks understanding of the principles underpinning their successes, drawing guidance from this to inform increasingly sustainable decision-making.

Though the book is inherently positive, Chap. 2, *Nature's sinking ark*, outlines some of the hard realities of the modern world and the prognoses for eroding natural capital. This is contextually important, as it informs both the need for urgent and committed action but also illustrates the roots of a variety of contemporary problems. We and nature share a conjoined destiny, tied as human prospects are to the ecosystems that gave us birth and continue to give us life from a basic biological sense through to supporting our economic activities and capacities to live fulfilled lives. Yet today, the metaphorical ark of nature is disintegrating under our collective weight, already vastly impoverished in diversity, resilience and capacities to support continuing human wellbeing. The concept of socio-ecological systems,[3,4,5] central to this understanding, defines the systemic interdependencies between people and other aspects of natural systems. Socio-ecological systems can be in a cycle of degradation, wherein declining ecosystem health and functioning inevitably limits human opportunity. Alternatively, a regenerative cycle can occur where protection or restoration of foundational ecosystem processes supports future human wellbeing. Oversight of the wider unintended ramifications of narrowly exploitative uses of supportive ecosystems, though otherwise laudable in their intent to efficiently meet our needs, very often drives 'degenerative landscapes' in which narrowly-framed uses inadvertently degrade ecosystems and, with them, interlinked human prospects.

Chapter 3, *Rebuilding the ark*, then embarks on a global journey, collating case studies from across the world to illustrate and characterise instances where people have collaborated to not merely halt damage to supporting ecosystems but to restore them as a foundation for reversing cycles of degradation. These 'regenerative landscapes'—recovering or protected ecosystems supporting socio-economic benefits and expanding opportunities for all—contain vital lessons informing a change of course to realise a sustainable future, not merely in a biological sense but one that also embodies human security, opportunity, prosperity, beauty and the achievement of life potential.

Chapter 4, *Our conjoined future*, draws upon 'regenerative landscape' case studies to emphasise the need for a fully systemic approach to solutions that optimise outcomes for all ecosystem services and their associated beneficiaries, rather than perpetuating current narrow exploitation patterns that favour the few yet overlook the many. The chapter also addresses what sustainable development means in a world of growing human numbers reliant upon declining natural resources. It emphasises the importance of going well beyond current minimalist understandings and regulatory transpositions of 'lightening society's loads on ecosystems',

[3] Holling, C.S. (2001). Understanding the complexity of economic, ecological, and social systems. *Ecosystems*, 4(5), pp. 390–405.

[4] Glaser, M., Krause, G., Ratter, B. and Welp, M. (2008). Human/nature interaction in the anthropocene: Potential of social-ecological systems analysis. *GAIA*, 17(1), pp. 77–80.

[5] Madrid-Lopez, C. and Giampietro, M. (2015). The water metabolism of socio-ecological systems reflections and a conceptual framework. *Journal of Industrial Ecology*, 19(5), pp. 853–865.

instead realising the explicit bold, intergenerational vision of the 1987 'Brundtland Commission' definition of sustainable development. Creating regenerative socio-ecological cycles is essential to secure future human security and opportunity, essentially reclaiming the definition of sustainable development such that it is recognised as vastly more than some woolly, altruistic or 'nice to have' aspiration. Systemic application of the STEEP framework—an acronym for constituent social, technological, economic, environmental and political elements—is introduced as a systems-based model for understanding social and ecological interdependencies within complex contexts of technology choice and management in political economies. Consideration is also given to where leadership happens in society.

Chapter 5, *A systemic decision-support framework*, establishes key and contextual questions related to each component of STEEP as a framework to guide decision-making at all scales on a systemically connected basis. This framework is proposed to inform decision-making from a systemic context in pursuit of genuinely sustainable, resource-conserving and equitable, regenerative socio-ecological systems, or at least to promote progress from current practice towards that ideal goal cognisant of shortfalls and areas for further research, reform and improvement. The framework has relevance to decision-making across all major societal policy areas, the chapter providing applied examples not only within policy areas but importantly of integration between them.

Chapter 6, *Epilogue: Rebuilding the ark—Yes, we can!* emphasises that this is all about culture change, addressed with the optimism that regenerative change is already happening in scattered instances across the world. These exemplars can teach us much about how to rethink our decisions at international, national, state and right down to highly localised scales, and to achieve coherence between all these scales. Action at all levels, and from all sectors of society, is required to rebuild the metaphorical ark of natural species, processes and ecosystem services supporting continuing human security and wellbeing.

The book is written with accessibility by a wider diversity of scientific, policy, activist, and informed and concerned public readers in mind. Consequently, the extensive and robust body of published scientific and other sources underpinning examples and arguments presented throughout feature as footnotes, so as not unnecessarily to disrupt the flow of the text. For that same purpose, many case studies and illustrative examples also appear in Boxes. Many of the more prominent examples are from India and elsewhere in South Asia, where a great deal of my recent work has been focused, but the wider case studies within the book encompass all continents and nations, spanning a range of stages of development. The principles derived are generically valid.

As well as to inform, a primary purpose of this book is to inspire a 'can do' mind set, empowering readers to play their parts in tackling the pressing sustainability challenges facing the world today, and to rebuild the foundations of enhanced human prospects for the future.

2

Nature's Sinking Ark

As related in the Introduction, the purpose of this book is to focus on success stories, solutions and reasons to be optimistic. But it would be remiss to overlook the hard realities giving context and urgency to this need to look for brighter lights in a darkening and severely challenged world. In essence, the metaphorical ark of nature is disintegrating under our collective societal weight, and is today vastly impoverished in its diversity, resilience and capacities to keep afloat prospects for continuing human wellbeing. The very real prognosis is that, without informed and committed interventions, we too will sink along with our supporting vessel, for which we have no alternative.

2.1   The State of Nature

Recent human history has not treated the natural world kindly. Evidence from the 2005 Millennium Ecosystem Assessment[1] (MA), a global assessment called for by the United Nations' then Secretary-General Kofi

---

[1] Millennium Ecosystem Assessment. (2005). *Ecosystems and Human Well-being*. Island Press.

9
M. Everard, *Rebuilding the Earth*, https://doi.org/10.1007/978-3-030-33024-8_2

Annan in 2000 with the objective of assessing the consequences of ecosystem change for human well-being and a scientific basis for their sustainable use, provided a state-of-the-art scientific appraisal of the condition and trends in the world's ecosystems. Importantly, it also addressed the 'ecosystem services'—the many and diverse benefits flowing to humanity from nature—that these natural systems provide, including for example clean water and air, food and forest products, flood control, recreational and spiritual resources, and processes underpinning the resilience and continued functioning of the ecosystems themselves. Amongst many of the general conclusions of the MA about trends in major global habitat types were that the number of species on the planet is steeply declining and also homogenising through invasive introductions. Humans over the past century have increased extinction by up to 1,000 times over background rates. As just one example, 20% of the world's coral reefs were found to have been lost with an additional 20% degraded in the last decades of the twentieth century, with subsequent observations in 2017–2018 by the Australian Institute of Marine Science[2] finding that coral cover on the Great Barrier Reef has continued to steeply decline at a rate that has not been observed in the historical record. The MA also found that the atmospheric concentration of carbon dioxide had increased from 280 to 376 parts per million (ppm) between 1750 and 2003 primarily due to combustion of fossil fuels and land use changes, 60% of that increase occurring since 1959. There has been a subsequent continuing upwards trend reaching 408.02 ppm as of November 2018.[3]

Practices that we accept as normal and acceptable can be prominent drivers of declining ecological capacity. As one example, current intensive agricultural practices represent a major threat to biodiversity and ecosystem services. By 2050, intensive farming could account for 70% of the predicted loss of terrestrial biodiversity,[4] and the United Nations[5] also

---

[2] Australian Institute of Marine Science. (2018). *Long-term Reef Monitoring Program—Annual Summary Report on Coral Reef Condition for 2017/18.* Australian Institute of Marine Science, Townsville. https://www.aims.gov.au/reef-monitoring/gbr-condition-summary-2017-2018, accessed 31 December 2018.

[3] https://www.co2.earth/, accessed 31 December 2018.

[4] UN Water. (2017). *Wastewater the Untapped Resource.* The United Nations World Water Development Report.

reports on multiple desertification effects. 52% of drylands used for agriculture are moderately or severely affected by soil degradation, affecting 1.5 billion people globally with serious implications for hunger now and into the future. Across this range, an estimated 27,000 associated species are lost each year. Arable land loss is emerging as a major global problem, occurring at 30–35 times the historical rate with 24 billion tons of fertile soil (one of the most significant, non-renewable geo-resources) eroded from global landscapes annually.

Tropical forests also continue to disappear at a rapid and accelerating rate.[6] Global forest area declined by approximately 40% in the last three centuries, three-quarters of this during the last two centuries.[7] Forests have completely disappeared in 25 countries, with greater than 90% loss of forest cover in another 29 countries. Clearance of natural forests in the tropics continues at 13 million hectares per year[8] (approximately the size of Greece), with ongoing degradation and fragmentation of many remaining forests. At least half of recent global deforestation is caused by demands for land to serve commercial agriculture, with 49% of tropical deforestation between 2002 and 2012 due to illegal conversion.[9] Forest loss remobilises vast reserves of stored carbon from biomass and soil, generating nearly 50% more greenhouse gases than the global transportation

---

[5] United Nations. (n.d.). *Desertification.* United Nations. http://www.un.org/en/events/desertificationday/background.shtml, accessed 27 May 2016.

[6] Hansen, M.C., Potapov, P.V., Moore, R., Hancher, R., Turubanova, S.A., Tyukavina, A., Thau, D., Stehman, S.V., Goetz, S.J., Loveland, T.R., Kommareddy, A., Egorov, A., Chini, L., Justice, C.O. and Townshend, J.R.G. (2013). High-resolution global maps of 21st-century forest cover change. *Science*, 342(6160), pp. 850–853.

[7] Shvidenko, A., Barber, C.V., Persson, R., Gonzalez, P., Hassan, R., Lakyda, P., McCallum, I., Nilsson, S., Pulhin, J., van Rosenburg, B. and Scholes, R. (2005). Forest and woodland systems. In: *Ecosystems and Human Well-being: Current State and Trends*. Millennium Ecosystem Assessment. Chap. 21, pp. 585–621. http://www.millenniumassessment.org/documents/document.290.aspx. pdf, accessed 7 March 2016.

[8] UN FAO. (2011). *Payments for Ecosystem Services and Food Security*. United Nations Food and Agriculture Organization, Rome. 300pp.

[9] Lawson, S., Blundell, A., Cabarle, B., Basik, N., Jenkins, M. and Canby, K. (2014). *Consumer Goods and Deforestation: An Analysis of the Extent and Nature of Illegality in Forest Conversion for Agriculture and Timber Plantations*. Forest Trends Report Series: Forest Trade and Finance. http://www.forest-trends.org/documents/files/doc_4718.pdf, accessed 2 March 2016.

sector.[10] Deforestation also destroys habitats for a diversity of species, and degrades water storage and purification, natural buffering of storm energy and protection from erosion.

In the world's oceans, 7% of 600 marine fisheries monitored in 2005 were in a depleted state, with 17% over-exploited, 52% fully exploited, and a mere 1% recovering from depletion.[11] Fish stocks in some tropical areas are predicted to decline by an estimated 40–60% due to climate change alone,[12] with significant implications for the food security and livelihoods of many millions of people in developing coastal states. The estimated USD$83 billion losses (more than 110% of the GDP of Kenya) in annual global revenue caused by overfishing[13] is also leading to a substantial leakage of value out of the ocean economy (Fig. 2.1).

The May 2019 Global Assessment Report[14] of the Intergovernmental Science-Policy Platform on Biodiversity and Ecosystem Services (IPBES) documented an unprecedented dangerous decline in global biodiversity with approximately 75% of the land-based environment and 66% of the marine environment 'severely altered' by human actions. An accelerating rate of species extinction was also reported, roughly tens to hundreds of times greater than natural extinction rates, putting at risk about 1 million amongst the estimated 8 million species on Earth.

Loss of biodiversity is proceeding at such a rate that we are facing a mass extinction event,[15] an irreversible loss to the planet that also threat-

---

[10] Nabuurs, G.J., Masera, O., et al. (2014). Chapter 9: Forestry. In: *IPCC Fifth Annual Assessment Report—Climate Change 2014: Impacts, Adaptation, and Vulnerability.* https://www.ipcc.ch/pdf/assessment-report/ar4/wg3/ar4-wg3-chapter9.pdf, accessed 2 March 2016.

[11] FAO. (2005). Review of the State of World Marine Fisheries Resources. FAO Fisheries Technical Paper 457. Food and Agriculture Organisation of the United Nations, Rome. http://www.fao.org/3/y5852e/Y5852E00.pdf, accessed 2 March 2016.

[12] Intergovernmental Panel on Climate Change (IPCC). (2014). *Climate Change 2014: Impacts, Adaptation and Vulnerability. Part A: Global and Sectoral Aspects.* Cambridge University Press, UK. www.ipcc.ch/report/ar5/wg2/, accessed 14 February 2019.

[13] The World Bank. (2015). *The Sunken Billions Revisited: Progress and Challenges in Global Marine Fisheries.* http://www.worldbank.org/en/topic/environment/brief/the-sunken-billions-revisited-progress-and-challenges-in-global-marine-fisheries, accessed 14 February 2019.

[14] Brondizio, E.S., Settele, J., Díaz, S. and Ngo, H.T. (2019). *Global Assessment on Biodiversity and Ecosystem Services of the Intergovernmental Science-Policy Platform on Biodiversity and Ecosystem Services (IPBES).* IPBES. https://www.ipbes.net/global-assessment-biodiversity-ecosystem-services.

[15] Barnosky, A.D., Matzke, N., Tomiya, S., Wogan, G.O.U., Swartz, B., Quental, T.B., Marshall, C., McGuire, J.L., Lindsey, E.L., Maguire, K.C., Mersey, B. and Ferrer, E. (2011). Has the Earth's sixth mass extinction already arrived? *Nature*, 471(7336), pp. 51–57.

Fig. 2.1 Human demands on marine resources, and the means applied to service them, are placing marine ecosystems under increasing stress. (Image © Dr Mark Everard)

ens humanity's life support system.[16,17,18] Beyond the sheer loss of numbers of species, we are also seeing unprecedented 'biotic homogenisation',

---

[16] Díaz, S., Fargione, J., Chapin, F.S. and Tilman, D. (2006). Biodiversity loss threatens human well-being. *PLoS Biology*, 4(8), pp. 1300–1305.

[17] Cardinale, B.J., Duffy, J.E., Gonzalez, A., Hooper, D.U., Perrings, C., Venail, P., Narwani, A., Mace, G.M., Tilman, D., Wardle, D., Kinzig, A.P., Daily, G.C., Loreau, M. and Grace, J.B. (2012). Biodiversity loss and its impact on humanity. *Nature*, 486(7401), pp. 0–9.

[18] Hooper, D.U., Adair, E.C., Cardinale, B.J., Byrnes, J.E.K., Hungate, B. A., Matulich, K.L., Gonzalez, A., Duffy, J.E., Gamfeldt, L. and O'Connor, M.I. (2012). A global synthesis reveals

in which the introduction of non-native species and the extinction of local biodiversity increases the genetic, taxonomic or functional similarity of different locations.[19] The net result is simplified, less locally adapted and increasingly vulnerable ecosystems producing a narrower set of ecosystem services supporting the needs of society.

2.2 Nature and People

All ecological degradation has major consequences for people, indivisibly linked as we are in tight socio-ecological cycles for all of our biological and economic needs and higher aspirations. Everything that we do affects this integrated system, just as every change in the natural system has multiple ramifications for us.

Interdependence between human and ecological systems underpins the concept of ecosystem services, defined by the MA[20] as "...*the benefits that people derive from nature*" spanning a broad diversity of fundamental and varied benefits to society. Though by definition anthropocentric, framed by the diverse dimensions of human wellbeing supported by natural systems, awareness of the interlinked suite of ecosystem services informs us of the substantial and multiple values conferred to us by natural systems. A globally consistent classification of ecosystem services derived by the MA (Box 2.1) has been variously revised, but virtually all derivative classifications highlight the multiplicity of services provided to society by nature. Many of these services, such as valued landscapes, air purification and soil formation, are not traded in markets and are commonly omitted from regulatory requirements and overlooked in ecosystem use and management. Yet, as ecosystem services are produced as an integrated set, all

---

biodiversity loss as a major driver of ecosystem change. *Nature*, 486(7401), pp. 105–108.

[19] McKinney, M.L. and Lockwood, J.L. (1999). Biotic homogenization: A few winners replacing many losers in the next mass extinction. *Trends in Ecology and Evolution*, 14, pp. 450–453.

[20] Millennium Ecosystem Assessment. (2005). *Ecosystems and Human Well-being: General Synthesis.* Island Press, Washington, DC. http://www.millenniumassessment.org/en/Synthesis.html, accessed 15 March 2016.

uses and manipulations of ecosystems affect the functioning of the system as a whole and with it the production of all linked services.

Box 2.1 Ecosystem Services Classification of the Millennium Ecosystem Assessment

The Millennium Ecosystem Assessment framework recognises four qualitatively different categories of ecosystem services:

- Provisioning services can be extracted for human uses, including food, fibre, fresh water, biochemical substances and energy;
- Regulating services moderate, for example, flows and quality of air and water, and natural moderation of erosion, diseases, climate and pollination;
- Cultural services comprise non-material benefits such as spiritual enrichment and educational, tourism and recreational opportunities; and
- Supporting services include soil formation, oxygen generation, primary production, cycling of nutrients and water, and habitat for wildlife. Though redefined in some subsequent classifications as processes rather than directly exploited services,[21,22] supporting services remain important elements to factor into policy and management to protect the characteristics, integrity, functioning and resilience of ecosystems and their capacities to supply other more directly utilised services.

One model that departs from this general structure is that of the IPBES, which structures the values of nature into three dimensions of 'Intrinsic value', 'Instrumental value' and 'Relational value' (Box 2.2). The IPBES values model mines deeper into the inherent values of nature, in addition to expanding upon the differing ways in which humanity benefits from them.

---

[21] Braat, L.C. and de Groot, R. (2012). The ecosystem services agenda: Bridging the worlds of natural science and economics, conservation and development, and public and private policy. *Ecosystem Services*, 1, pp. 4–15.

[22] TEEB. (2010). *The Economics of Ecosystems and Biodiversity: Mainstreaming the Economics of Nature: A Synthesis of the Approach, Conclusions and Recommendations of TEEB*. The Economics of Ecosystems and Biodiversity (TEEB). http://doc.teebweb.org/wp-content/uploads/Study%20 and%20Reports/Reports/Synthesis%20report/TEEB%20Synthesis%20Report%202010.pdf, accessed 25 September 2016.

> **Box 2.2 Three Broad Dimensions of Values of Nature Defined by IPBES[23]**
>
> - *Intrinsic value*: the inherent value of nature, independent of any human use or judgement. This spans individual organisms; biophysical assemblages (populations, communities, ecosystems, biomes, the biosphere, Gaia, etc.); biophysical processes; and biodiversity (genetic, functional, taxonomic and phylogenetic diversity, uniqueness, etc.)
> - *Instrumental value*: contributions of nature to the achievement of human quality of life. This includes many ecosystem services recognised in other classifications but also wider considerations spanning: the biosphere's ability to enable human endeavour; nature's ability to supply benefits to humanity; and 'nature's gifts, goods and services' (including various regulating, provisioning and cultural ecosystem services).
> - *Relational value*: contributions to good human quality of life, including desirable relationships among people and between people and nature. This spans various dimensions including: security and livelihoods; sustainability and resilience; diversity and options; living well and in harmony with nature and Mother Earth; health and wellbeing; education and knowledge; inspiration; identity and autonomy; good social relations; art and cultural heritage; spirituality and religions; and governance and justice.

Given the inherently interlinked nature of ecosystem services and cumulative impacts upon them from any form of ecosystem change, the decline in global aquatic ecosystems, including the quantity and quality of water within them, is of far more than altruistic concern. There has been a tendency not only to make presumptions in favour of technically efficient water extraction and transfer technologies but to look increasingly remotely for perceived surplus water resources that in turn become depleted, marginalising communities in the places from which water is withdrawn. This is a potential, and in many localities an actual, cause of civil unrest,[24] degrading supporting ecosystems in an unsustainable model characterized as a "civil engineering paradigm" in which this engineering-based approach to meet the mushrooming water demands of growing cities embarks on a cycle of "taking more from further".[25] The reality is that dense human settlements do need 'heavy engi-

---

[23] IPBES. (2015). *Preliminary Guide Regarding Diverse Conceptualization of Multiple Values of Nature and Its Benefits, Including Biodiversity and Ecosystem Functions and Services (Deliverable 3(d))*. IPBES/4/INF/13. https://www.ipbes.net/sites/default/files/downloads/IPBES-4-INF-13_EN.pdf.

[24] Birkenholtz, T. (2016). Dispossessing irrigators: Water grabbing, supply-side growth and farmer resistance in India. *Geoforum, 69*, pp. 94–105.

[25] Barraqué, B., Formiga Johnsson, R.M. and Nogueira de Paiva Britto, A.L. (2008). The development of water services and their interaction with water resources in European and Brazilian cities. *Hydrology and Earth System Science, 12*, pp. 1153–1164, p. 1156.

neering' solutions to address concentrated demands for both clean water supply and wastewater treatment. However, sustainable solutions lie not so much in reversion to either engineered or nature-based approaches alone, but in their context-specific hybridisation supporting local, rural needs whilst replenishing ecosystems from which large-scale water resources are withdrawn. The multiple dimensions of current problems of degradation of linked ecosystem functioning and human wellbeing are exemplified by transitions in water management in contemporary India, threatening the viability of water resources and so driving a degrading socio-ecological cycle (Box 2.3).

Box 2.3 Unsustainable Transitions in Water Management in India

Transitions in water management in India during the twentieth century demonstrate how changes in technology and governance can drive negative unintended consequences for the viability of landscapes and inextricably linked socio-economic welfare. This situation is replicated throughout much of the drier parts of the developing world.

Most of central India's rain falls during a short monsoon period, with the hot climate driving high evaporation throughout the rest of the year. This results in substantial dependence on groundwater for year-round access. Indian communities have consequently developed a huge variety of locally geographically and culturally attuned water-harvesting structures (WHSs) over centuries and millennia. WHSs embody traditional knowledge and community-based collaboration, intercepting monsoon run-off to recharge groundwater protected from high evaporative rates, and consequently accessible throughout the year.

Contemporary political, technological and economic pressures are combining to break down the community collaboration essential for WHS operation and the sharing of sustainably managed resources. Pervasion of mechanised tube wells tap deeper, often receding and now increasingly contaminated groundwater on a non-renewable and competitive basis. This impact is exacerbated by a policy environment conflating increased irrigation with a narrow model of economic growth that includes subsidised energy for pumping. As people compete rather than collaborate, the community basis underpinning historic sustainable water use and sharing also degrades. Degradation of water resources and linked ecosystems drives aridification and consequent farmland and village abandonment across much of central India in a negative socio-ecological cycle.[26]

---

[26] Everard, M. (2015). Community-based groundwater and ecosystem restoration in semi-arid north Rajasthan (1): Socio-economic progress and lessons for groundwater-dependent areas. *Ecosystem Services*, 16, pp. 125–135.

The 2019 IPBES Global Assessment Report[27] backs up its alarming conclusions about the rapid and accelerating deterioration of global ecosystems with the conclusion that, through the negative outcomes of our actions on ecosystems, we are eroding the foundations of our economies, livelihoods, food security, health and quality of life worldwide.

2.3   People and Nature

People and wildlife are net victims, but it is people who are the net causes of this ecological mayhem. This is not by intent, but largely through the narrow vision and framing of our patterns of ecosystem use.

Through our growing numbers and the reach of modern energised technologies, humanity is no mere marginal addition to the workings of planetary systems; we have become a dominating influence. The Holocene epoch was defined by natural forces dating back to the end of the last major 'ice age' of the Pleistocene 11,700 years ago. Many now recognise a new geological age defined as the Anthropocene, with humans now constituting a dominant influence shaping Earth's ecosystems.[28] Though this redefinition is contested, it is true that human activities exploit and influence ecosystems reciprocally and profoundly. Yet this tight interdependence between humanity and the functioning of ecosystems is only poorly internalised by regulations, markets, fixed assumptions and vested interests. A new paradigm of development is required founded on the integrated workings of whole socio-ecological systems.

As one pervasive global example, policy and market instruments reward farmers for the efficient intensive production of cheap food and other commodities, with token or else no rewards to recognise and maintain the many other ecosystem services that landscapes provide and from which people benefit. For example, intensive farming systems, powered by an influential agribusiness industry, tend to overlook impacts upon hydro-

---

[27] Brondizio, E.S., Settele, J., Díaz, S. and Ngo, H.T. (2019). *Global Assessment on Biodiversity and Ecosystem Services of the Intergovernmental Science-Policy Platform on Biodiversity and Ecosystem Services (IPBES)*. IPBES. https://www.ipbes.net/global-assessment-biodiversity-ecosystem-services.

[28] Crutzen, P.J. and Stoermer, E.F. (2000). The 'anthropocene'. *Global Change Newsletter*, 41, pp. 17–18.

logical processes buffering droughts and floods, purification of water and air, aesthetic and educational places, carbon storage and habitat for wildlife including fish and other constituents of recreational value. These intensive systems even substantially overlook the protection of fertile soils—the primary resource of continuing agricultural productivity—from degradation, erosion and loss. Similar trends are observed, as further examples, in the narrow reward systems for production of forest and marine fishery products, loss of habitats for urban, industrial and infrastructure development, mining of economically valued resources, and the consequences of substantial releases of contaminants into the atmosphere and water systems.

The 'carrying capacity' of ecosystems to withstand narrow forms of exploitation is often overlooked in terms of longer-term consequences for the wealth of natural services upon which our future prospects depend. Cumulatively, narrowly focused exploitation habits tend to degrade the integrity, functioning and supportive capacities of marine systems, forests, rangelands and other major habitat types globally. Population growth, climate instability and globalised supply chains increase their impact.[29] Habitat exploitation for narrow benefits tends to create 'degenerative landscapes', locked into a spiral of closely linked ecological, ecosystem service and socio-economic decline. The demands of contemporary global society effectively consume 1.5 'Planet Earths'.[30] As we only have one Earth, our singular essential resource to meet the demands of a growing human population, there is a pressing need to use landscapes and other ecosystems in more sustainable and integrated ways if the prognosis of non-renewable, degenerative landscape exploitation is to be halted and reversed to sustain continuing human security and opportunity.

Furthermore, competition for limited resources underpins conflicts globally. India, as one example amongst many other countries, has experienced many instances of deaths resulting from civil uprisings against dam and water transfer schemes perceived as dispossessing people of their rights.[31] A more secure and functional natural resource base—water,

[29] Millennium Ecosystem Assessment. (2005). *Ecosystems and Human Well-being*. Island Press.
[30] Global Footprint Network. (2016). *Living Planet Index 2016*. Global Footprint Network. http://www.footprintnetwork.org/en/index.php/GFN/, accessed 14 November 2016.
[31] World Commission on Dams. (2000). *Dams and Development: A New Framework for Better Decision-making*. Earthscan, London.

food, soil, energy, timber and other resources—represents a major contribution to reducing potential for conflict.

2.4   Consequences and Challenges

Global humanity made significant, if regionally variable, social and economic progress during the twentieth century. We have witnessed dramatic and increasing technological advances to enhance the power with which we have been able to exploit ecosystems. However, these activities, undertaken to achieve laudable and desirable outcomes such as affordable food, construction materials, fuel and other commodities, as well as places to live and work, have generally focused on narrow ends to the detriment of whole-system functioning and resilience, and the production of the breadth of services vital to underpin a secure and fulfilled future.

Today's world is consequently full of non-systemic outcomes, often driven by good intentions but generating a diversity of externalities (positive or, as is common with ecosystem uses, negative consequences affecting other services and their beneficiaries). A cross section of examples is summarised in Box 2.4.

Box 2.4 Unintended Negative Outcomes Arising from Narrow Exploitation of Ecosystems

- America's disastrous 'Dust Bowl' resulted from economic stimuli in response to the US Great Depression. Tracts of formerly vegetated prairie were released for farming. However, removal of protective prairie vegetation resulted in massive erosion, exacerbated by a severe drought and deeper ploughing technology. As wind and rain whipped away soil, fertile lands across the American Midwest were converted into dry dust clouds swamping farmsteads and rural towns, displacing hundreds of thousands of people.[32] Approximately 3.5 million people moved out of the Plains states between the 1930s and 1940s,[33] driven by rapid breakdown of the unaccounted resilience, ecology and socio-economic benefits of prairie landscapes.

---

[32] Hakim, J. (1995). *A History of Us: War, Peace and All that Jazz*. Oxford University Press, New York.
[33] Worster, D. (1979). *Dust Bowl: The Southern Plains in the 1930s*. Oxford University Press.

- Industrialisation of marine capture fisheries through increasingly powerful engines enable fleets to access more remote and deeper waters, using increasingly sophisticated GPS and sonar technology to target remaining shoals of fish. In the absence of effective fishery management policies, this has resulted in the collapse or degradation of formerly rich fish stocks.[34,35]
- Forests provide valuable construction, fuel, medicinal and other benefits. However, if extraction exceeds regeneration, or unsustainable methods are used, forests lose their resilience and capacities to produce a diversity of linked ecosystem services. Some of these beneficial services include regulation of water quality and quantity, provision of food, fuel and fibre, genetic, medicinal and ornamental resources, climate regulation, recreational potential and a wide range of other regulatory, cultural and supporting services.[36]
- Large dams impose significant 'take' of land and other natural resources. Where landscapes were formerly in common stewardship, dispossession, disempowerment and displacement of landless people is common.[37] Large dams also fundamentally change riverine ecosystems, including halting regeneration of soil fertility on downstream floodplains, perturbed fish stocks and proliferation of water-borne diseases due to simplified hydrology. All of these impacts compromise the ability of people to sustain livelihoods, skewing benefits to narrow sectors of recipients of piped water and hydropower.
- In the UK, the *State of Nature 2016* report, published in collaboration between a wide range of UK conservation and research organisations, found that 56% of species declined, with 40% showing strong or moderate declines (including 60% of vascular plants, 62% of butterflies and 49% of bird species) between 1970 and 2013.[38] Nature is faring worse in the UK than in most other countries with 1 in 10 wildlife species facing extinction due to the destructive impact of intensive farming, urbanisation and climate change on plants, animals and habitats.

---

[34] Kurlansky, M. (1999). *Cod: A Biography of the Fish that Changed the World.* Penguin Books, London.

[35] Kurlansky, M. (2008). *The Last Fish Tale: The Fate of the Atlantic and Our Disappearing Fisheries.* Jonathan Cape.

[36] Everard, M., Longhurst, J.W.S., Pontin, J., Stephenson, W. and Brooks, J. (2017). Developed-developing world partnerships for sustainable development (1): An ecosystem services perspective. *Ecosystem Services,* 24, pp. 241–252.

[37] World Commission on Dams. (2000). *Dams and Development: A New Framework for Better Decision-making.* Earthscan, London.

[38] RSPB and others. (2016). *State of Nature 2016.* https://www.rspb.org.uk/globalassets/downloads/documents/conservation-projects/state-of-nature/state-of-nature-uk-report-2016.pdf, accessed 8 May 2019.

- Even ostensibly pro-environmental measures such as establishment of wildlife reserves may produce negative outcomes where wider ramifications are overlooked. Establishment of the Kruger National Park in South Africa in 1898 resulted in a world-class National Park, but was achieved through mass clearances of tribal people who had lived on and made use of the productivity of the land for generations. Many descendants still live in abject poverty in settlement camps. This pattern is repeated in the USA where Native Americans were expelled from land designated as National Parks.[39] Establishment of Yellowstone National Park from 1872 saw hundreds of Native Americans evicted or killed, the authorities of the time also reneging on treaty promises.[40]

When wider systemic impacts are overlooked, it is all but inevitable that the integrity of supporting ecosystems will degrade (Fig. 2.2). For all our best intentions, human pressures resulting from means implemented to serve rapidly growing demands for food, fresh water, timber, fibre and fuel have resulted in and continue to drive serious and continuing degradation of most global major habitat types. Ecosystem damage over the last 50 years of the twentieth century was greater than in any comparable period of human history.[41] The *Global Risks Report 2019*[42] of the World Economic Forum included six ecosystem-related factors—extreme weather events, failure of climate change mitigation and adaptation, major natural disasters, man-made environmental damage, major biodiversity loss and ecosystem collapse, and water crises—amongst its top ten risks by likelihood over the next 10 years.

Major distributional concerns arise from unequal sharing of benefits and costs to society at large, including future generations, when integrating risks. Examples are myriad, including the unintended consequences of contemporary farming for downstream flooding and drought, costs of

[39] Merchant, C. (2007). *American Environmental History: An Introduction*. Columbia University Press, New York.

[40] Spence, M.D. (1999). *Dispossessing the Wilderness: Indian Removal and the Making of National Parks*. Oxford University Press, Oxford.

[41] Millennium Ecosystem Assessment. (2005). *Ecosystems and Human Well-being*. Island Press.

[42] World Economic Forum (WEF). (2015). *The Global Risks Report 2019*. World Economic Forum (WEF), Davos. https://www.weforum.org/reports/the-global-risks-report-2019/, accessed 14 February 2019.

Fig. 2.2   Large dams such as the Indira Gandhi Sagar scheme on the Narmada River in Madhya Pradesh, India, are efficient in storing and diverting large-scale water resources and producing hydropower, but degrade river ecosystems and their many services to a wide range of beneficiaries. (Image © Dr Mark Everard)

treating more contaminated water, siltation of rivers, declining fish stocks and other wildlife, loss of amenity, and other factors co-produced through myopic maximisation of the single benefit of farmed commodity produce.

2.5   Misconceptions of Value

As noted above, we tend to define infrastructure and value in narrowly anthropocentric terms, omitting the foundational role and value of natural systems that support our wellbeing and opportunities now and, we hope, into the future.

Whilst developed world nations draw resources from far beyond 'home' landscapes, with increasing dependence on globalised supply chains to support dense populations inhabiting highly modified landscapes, the fact remains that natural systems remain the ultimate 'suppliers' and 'trash collectors'. And, whilst we may rest on historic trading advantages entrenched in the world's political and economic power asymmetries in order to substitute for the depletion of domestic resources, the hard reality remains that our biophysical needs as well as the metabolism of our economy ultimately rests upon natural infrastructure, however remotely.

A gross and pervasive example here amongst many is the 'defence' from flooding of housing, farmland, commercial properties, transport lines and other structures built on floodplains. It should be self-evident from the appearance of the word 'flood' in their name that these flat, riparian lands tend to become inundated during periods of heavy precipitation. However, industrialised society has regarded, and perplexingly continues to regard, these areas as nicely flattened places to site buildings, transport routes and to farm fertile, alluvial soils in ways that fail to account for changing water levels necessitating mechanical defences from the perceived depredations of nature. I remain baffled as to why flooding provokes such surprise each year, when pretty much every attentive student understands not only that floodplains typically fill with rising water on average every second year, but also that it is these very occurrences that sculpt the landscape. The clue, as we observed, is in the name. This stems from a misconception, or perhaps the word 'omission' is more appropriate, of the value of natural systems as compared to the things that we construct or exploit.

As one stark example, the market identifies a building on a floodplain as a capital asset, and the construction of a wall around it as a necessary service to protect that asset from inundation. From a more dispassionate, science-based view, it is the floodplain that is the natural capital, providing a multiplicity of services of substantial societal benefit. These benefits include storing floodwater and dissipating its energy, buffering flows between dry and wet seasons, providing habitat for wildlife including nursery areas for juvenile fishes, sequestering carbon and cycling nutrients, constituting characteristic and recreational landscapes, and many more besides. The dominant market model—perhaps the most globally prevalent and entrenched ideology in human history—is narrowly fixated on what is built, and the generally private property values attached to it. The objective reality is that nature and its processes respect no such largely arbitrary, anthropocentric lines on maps denoting property or territory, but adapt to changing conditions to deliver an abundance of eco-system services of great, if today largely disregarded and underappreciated, benefits to societal needs now and into the future. Regrettably, the short-term, profit-taking model dominates societal decisions and management responses to oddly misconceived 'natural disasters', such as flood damage to property built where flooding is an inevitability. Myopic exploitation erodes the multiplicity of other societal benefits that flow from the natural functioning of these landscape units.

Resource use habits and assumptions in the developed world, and increasingly also developing nations, are founded substantially on use and management approaches to achieve narrow outcomes whilst over-looking wider ramifications. We see this across the spectrum of human activities including in food production from land and oceans, fibre from forests, mined substances from underground, and so on. Yet we have knowledge of the fact that all interventions in ecosystems have systemic consequences for the ecosystem itself and for the many human benefits it provides (ecosystem services). Without this systemic perspective, honourable intentions to use ecosystems to support singular human needs may create cumulative negative impacts on ecosystem integrity and resilience, hence eroding overall value to and longer-term security for society.

2.6   Degenerative Landscapes

As most aspects of markets, government departments and legacy regulations and engineering solutions today focus on and reward narrow outputs from ecosystem use, many societal resource use habits effectively 'mine' terrestrial and aquatic ecosystems for short-term benefits. The term 'mine' reflects both extraction of resources exceeding natural regenerative rates, but also blindness to wider ramifications for all ecosystem services, their diverse beneficiaries and hence net societal value. A net long-term consequence of this myopic approach to resource exploitation in the driving of 'degenerative landscapes' (including waterscapes): ecosystems experiencing linked socio-ecological degradation that are all too common today (Fig. 2.3).

Fig. 2.3 Unsympathetic forestry practices, including plantations displacing native ecosystems or poor forest management such as clear felling that leads to soil erosion, can yield narrowly-framed short-term financial gains for a minority but tend to degrade biodiversity, ecosystem functioning and their many contributions to human security and opportunity. (Image © Dr Mark Everard)

There is no inherent problem with using landscapes and waterscapes to support human needs. However, we now face unprecedented issues of scale, in terms of our large and still burgeoning population and our tech-nological 'reach' to exploit productive environments for the sole advan-tage of our species, substantially compounded by a narrow focus on maximisation of single or narrow benefits and blindness to other linked ecosystem services. The consequent pervasive undermining of ecosystem integrity, and the distribution of benefits and costs enjoyed or imposed on service beneficiaries across society and time, defines many contempo-rary sustainability challenges.

2.7   We Can Work It Out

The challenge for us all facing a future of burgeoning human numbers and demands met by currently fast-declining ecosystems is to look broader, to interact with these crucial life support systems in ways that respect their underlying functionality and breadth of beneficial services. We need to seek optimisation of all ecosystem outputs, rather than a short-term focus on maximisation of just a selected few. Through a broader view and optimisation of a wider range of ecosystem services, we can potentially achieve more equitable, net societally beneficial and resil-ient outcomes.

In Chap. 3 of this book, we turn our attention to a smorgasbord of global examples where this optimising approach has been achieved, addressing foundational natural systems and the spectrum of people ben-efitting from them up front in the decision-making process.

To end this short consideration of the 'sinking ark' on a note of optimism, we are not bereft of vision and consensus-building around some of the world's major and most daunting sustainability challenges. This is particularly so in recognition of transnational issues that require the collaboration of global society. Box 2.5 outlines a subset of the many agreements addressing regionally, nationally and internationally concerted recognition and responses to threats common to all of global humanity.

Box 2.5 Agreements Addressing Issues of Concerns Beyond National Borders

International cooperation around transboundary environmental concerns yielding positive outcomes for human security include:

- 145 nations have territory within transboundary drainage basins, some resulting in conflict-related events but many more the subject of unilateral, bilateral or multilateral declarations or conventions on water management.[43,44]
- International responses to elevated radioactive contamination of the atmosphere from atmospheric nuclear weapon tests culminated in the Limited Test Ban Treaty[45] prohibiting all above-ground test detonations.
- Concerns about transboundary acid rain drove international action culminating in adoption of the UNECE Convention on Long Range Transboundary Air Pollution in 1979.[46]
- Discovery of the Antarctic 'ozone hole' in 1985 mobilised the international community to fix this 'hole in the sky', formalised under the Montreal Protocol on Substances that Deplete the Ozone Layer.[47] Treaties stemming from the Montreal Protocol have become the most widely ratified in UN history.
- Concerns about emissions of problematic chemicals triggered the European Union to develop a range of Directives including the Dangerous Substances Directive (67/548/EEC), the Sulphur Dioxide and Suspended Particulates EU Directive (80/779/EEC) and the Large Combustion Plant Directive (88/609/EEC).
- International mobilisation around climate change includes formation of the Intergovernmental Panel on Climate Change[48] (IPCC) in 1988 to provide scientific assessments of risks, potential consequences and possible adaptation and mitigation measures.[49]

---

[43] United Nations Department of Public Information. (2006). *Ten Stories the World Should Hear More About: From Water Wars to Bridges of Cooperation—Exploring the Peace-Building Potential of a Shared Resource.* http://www.un.org/Pubs/chronicle/2006/issue2/0206p54.htm#Water.

[44] www.unesco.org/water/wwap/pccp/.

[45] Arms Control Association. (1963). *Limited Test Ban Treaty (LTBT).* Arms Control Association. https://www.armscontrol.org/treaties/limited-test-ban-treaty, accessed 20 February 2019.

[46] UNECE. (1979). Convention on Long-Range Transboundary Air Pollution. http://www.unece.org/env/lrtap, accessed 21 June 2014.

[47] UNDP. (n.d.). *Montreal Protocol.* UN Development Program, New York. https://www.undp.org/content/undp/en/home/sustainable-development/environment-and-natural-capital/montreal-protocol.html, accessed 20 February 2019.

[48] www.ipcc.ch.

[49] IPCC. (2006). Principles governing IPCC work. *Intergovernmental Panel on Climate Change*, 28 April 2006. http://www.ipcc.ch/pdf/ipcc-principles/ipcc-principles.pdf, accessed 15 November 2013.

- In 1971, the Ramsar Convention on Wetlands of International Importance was adopted in the Iranian city of Ramsar, coming into force in 1975 and since incorporating almost 90% of UN member states from all the world's geographic regions as 'Contracting Parties'.[50] The Convention is an inter-governmental treaty that provides a framework for conservation and 'wise use' of wetlands and their resources,[51] the definition of 'wise use' since agreed as synonymous with sustainable development (Fig. 2.4).
- Over past decades, numerous multilateral environmental agreements have been adopted by governments to address biodiversity loss, amongst many other environmental concerns, although biodiversity continues to be unsustainably used and lost at genetic, species and ecosystem levels. Key multilateral agreements relating to biodiversity include the Convention on Biological Diversity with its Strategic Plan for Biodiversity 2011–2020[52] and the associated Aichi Biodiversity Targets[53] organised around five Goals, as well as a World Database on Protected Areas.[54]
- A series of UN-driven, high-profile intergovernmental gatherings have progressively raised the profile of environmental and sustainable development challenges in the international community for uptake by signatory nations. These include the 1972 'Stockholm Conference on Man and the Environment', the 1981/2 'World Conservation Strategy', the 1992 Rio de Janeiro 'Earth Summit', the 2002 'World Summit on Sustainable Development' (WSSD) in Johannesburg, and the 2012 'United Nations Conference on Sustainable Development ('Rio+20') in Rio de Janeiro.

Notwithstanding a nationalist, protectionist tendency across Europe and America at the time of writing with some significant retraction on agreements, visionary leadership to reverse the erosion of communal nat-

[50] Ramsar Commission. (2019). *About the Ramsar Convention*. Ramsar Commission, Gland. https://www.ramsar.org/about-the-ramsar-convention, accessed 1 January 2019.

[51] Ramsar Convention Secretariat. (2010). *Wise Use of Wetlands: Concepts and Approaches for the Wise Use of Wetlands*. Ramsar Handbooks for the Wise Use of Wetlands, 4th ed., vol. 1. Ramsar Convention Secretariat, Gland, Switzerland. http://www.ramsar.org/sites/default/files/documents/library/hbk4-01.pdf, accessed 25 June 2016.

[52] Convention on Biological Diversity. (2019). *Strategic Plan for Biodiversity 2011–2020*. Convention on Biological Diversity. https://www.cbd.int/undb/media/factsheets/undb-factsheet-sp-en.pdf, accessed 2 January 2019.

[53] Convention on Biological Diversity. (2019). *Aichi Biodiversity Targets*. Convention on Biological Diversity. https://www.cbd.int/undb/media/factsheets/undb-factsheet-sp-en.pdf, accessed 2 January 2019.

[54] IUCN. (2019). *World Database on Protected Areas*. International Union for Conservation of Nature (IUCN), Gland. https://www.iucn.org/theme/protected-areas/our-work/world-database-protected-areas, accessed 2 January 2019.

ural resources upon which continuing human wellbeing depends requires the concerted action of the global community. This is essential to override short-term vested interests entrenched in established, competitive market norms. In the absence of such leadership and real, internationally connected change, we may simply be deploying our more powerful remote sensing and other capabilities to become the first global species to monitor and document its own extinction.

Fig. 2.4 Wetlands are the focus of the 1971 Ramsar Convention on Wetlands of International Importance, the world's first habitat-specific convention, recognising their importance for wildlife but also for the many benefits they provide to humanity. (Image © Dr Mark Everard)

3

Rebuilding the Ark

How then do we go about rebuilding the metaphorical ark of nature's intricate ecosystems that, though much abused and degraded, has buoyed us on the voyage of human history? It is clearly not possible to withhold from any and all activities influencing the fabric and course of this 'mother ship', as all we do is integral to its inherently sustainable cycles of matter and energy. The answer lies in learning from the 'birds and bees', and all other biological components, in terms of meshing symbiotically with nature's renewable cycles.

Just as the world is waking up to the long-term, broad-scale hazards inherent in its cavalier exploitation and disposal of 'single use plastics' for purely immediate benefit, so too must we remove our blinkers regarding the narrow, single-purpose use of all natural resources. Aware as we now are of the systemic nature of ecosystems, their diverse services enjoyed by humanity, and the integrated nature of whole socio-ecological systems, we must recognise that every intervention in ecosystems has system-wide ramifications. This challenges our historic and legacy habits of focusing on ecosystem usage for narrow and immediate gains, unheeding of inevitable systemic consequences.

© The Author(s) 2020
M. Everard, *Rebuilding the Earth*, https://doi.org/10.1007/978-3-030-33024-8_3

In practice, now and into the future, there will always be one or (more rarely) a few linked ecosystem services, such as food or water production, flood management or a policy priority such as biodiversity protection or amenity provision, that emerges as a principal factor driving decision-making about ecosystem use or interventions. Historically, this need has, as we have seen, tended to be addressed as a sole driver blind to wider systemic ramifications. However, the desired outcome can more usefully be considered as an 'anchor service', or in other words a metaphorical 'anchor' for decision-making around which management options and exploitation techniques can be considered in systemic terms, taking account of the potential for co-delivery of a range of linked ecosystem service benefits of optimal societal benefit.[1] Optimisation of overall societal benefits then may be achieved by solutions that are generally ecosystem-based, or at least working in sympathy with natural processes, seeking to optimise benefits derived from a linked suite of ecosystem services including maintenance of the functioning of the providing ecosystem. These management measures constitute 'systemic solutions', defined as "…*low-input technologies using natural processes to optimise benefits across the spectrum of ecosystem services and their beneficiaries*".[2] Systemic solutions recognised under this initial definition include wetlands, washlands and urban ecosystem-based technologies optimised to achieve multiple ecosystem service outcomes simultaneously generated by focusing not solely on narrow ends but upon the foundational ecosystem processes that provide them. The principles implicit in 'systemic solutions' are that all ecosystem services, along with the rights of beneficiaries to those services, are systemically considered in any decisions. Such an approach encourages the optimisation of net societal value from ecosystem services; the benefits are not skewed towards a favoured few at the cost of benefits to any other overlooked beneficiaries (including future generations). A systemic solutions strategy implies a transition towards a more participatory and collaborative approach seeking optimal

---

[1] Everard, M. (2014). Nature's marketplace. *The Environmentalist*, March 2014, pp. 21–23.

[2] Everard, M. and McInnes, R.J. (2013). Systemic solutions for multi-benefit water and environmental management. *The Science of the Total Environment*, 461(62), pp. 170–179.

and sustainable outcomes. This connected world view needs progressively to supersede the legacy patchwork of narrowly-framed technical, legal and fiscal 'fixes'. At the very worst, a more systemically informed approach helps recognise and avert, or accommodate, negative consequences for formerly overlooked ecosystem services. Governance systems embodying this connected approach are easier to recognise at local scale, such as traditional village governance arrangements that have been historically adapted to achieve sustainable, enduring outcomes where people live in close proximity to supportive ecosystems.

A systemic solutions approach is a more inclusive, multi-factored approach than the predominantly competitive exploitation patterns inherent in contemporary western markets. Consideration of the functioning of ecosystems, and the distribution of benefits and costs that use and management options produce, can help inform the achievement of greater net beneficial and sustainable solutions. A systemic solutions strategy thereby implies transition from a narrow disciplinary and personalised approach to benefit realisation, essentially competitive with others' needs, towards a more participatory and collaborative approach seeking optimal, net socially beneficial sustainable outcomes. This systemic approach to ecosystem management addresses the foundations of supporting ecosystems from which diverse societal benefits flow, potentially reversing degrading socio-ecological cycles. It is also found at the heart of the global exemplars of 'regenerative landscapes' reviewed in this section of the book.

3.1   Reanimating the Water Cycle

Many of the pressing challenges in the developing world, also relevant to increasingly resource-constrained areas of the developed world, relate in one way or another to degradation of the water cycle. Our recent history of water management has tended to focus on technically efficient extraction of water to service immediate human demands. This has rather too frequently occurred in the absence of consideration of the workings of the water cycle, and particularly the replenishment of resources from which water is abstracted. This oversight leads to systematically degrading ecosystem vitality and functioning and, with it, negative consequences across whole socio-ecological systems.

The propensity to narrowly-framed exploitation is particularly pressing in India where unsustainable water management is compounding natural scarcity. Most of central India's rain falls during a short monsoon period, resulting in increasing dependence on groundwater throughout the long, dry seasons of high evaporation. Groundwater supports over 85% of India's rural domestic water requirements, 50% of urban and industrial water needs, and nearly 55% of irrigation demand.[3] India has adapted to this situation through a rich, centuries-long tradition of localised geographically and culturally attuned water-harvesting structures (WHSs) and management practices based on traditional knowledge and community-based collaboration. These structures and practices intercept monsoon run-off, promoting the recharge of groundwater protected from high surface evaporative rates and from which water is accessible throughout the year. Just some of the diversity of locally adapted WHSs across India include *Baudis, Khatris, Kuhls, Taanka, Naula, Dongs, Garh, Johadi, Virdas, jheels, Kattas* and *Eris*.[4] A broad diversity of localised, geographically and culturally adapted water management solutions working with natural processes are found across the varied agro-environmental zones of Rajasthan state alone.[5]

Yet a range of technological, policy and economic developments in recent decades have broken down the community structures of collaboration that are essential for WHS operation and resource sharing. As one technological example, mechanised tube wells can tap into deeper, often receding groundwater on a non-renewable and competitive basis. Problems of increasing physical scarcity are compounded by water quality concerns, as deeper groundwater in many regions of India is geologically contaminated by substances including fluoride, salt and arsenic,

---

[3] Government of India. (2007). *Report of the Expert Group on "Groundwater Management and Ownership" submitted to Planning Commission, September 2007.* Government of India, Planning Commission, New Delhi.

[4] Pandey, D.N., Gupta, A.K. and Anderson, D.M. (2003). Rainwater harvesting as an adaptation to climate change. *Current Science*, 85(1), pp. 46–59.

[5] Sharma, O.P., Everard, M. and Pandey, D.N. (2018). *Wise Water Solutions in Rajasthan.* WaterHarvest/Water Wise Foundation, Udaipur, India.

compounded by anthropogenic pollution particularly from industry and cities as well as increasing uses of agrochemicals. There has also been a technocentric trend since the 1940s favouring the construction of large dam-and-transfer schemes to supply water to areas of high demand (cities, industry and large-scale irrigation), generally overlooking consequent impacts upon ecosystems and people in the catchments from which water is diverted. Technology choices are generally driven by a policy environment seeking to maximise water supply for urban economies, irrigation and other intensive uses in the immediate term, without regard for the need to balance resource use with recharge. This approach has broken down the community basis underpinning historic sustainable water use and sharing, leading to abandonment of WHSs and consequently degrading water resources and other linked ecosystem services. The net consequence has been increased aridification, driving farmland and village outmigration and abandonment across significant areas of central India.

However, a broad range of initiatives are working to reverse these all too common cycles of linked ecological and socio-economic degeneration across India, as well as in other dryland regions of the developing world.

3.1.1 Reanimating Water and Livelihoods in Alwar District, Rajasthan

Rural outmigration and village abandonment consequent from over-extracted and degrading water systems are regrettably common across the desert and semi-arid state of Rajasthan. However, glowing examples of collaboration are also to be found there, where people have reinstituted water stewardship practices, reversing former cycles of degradation to rebuild a sustainable basis for regeneration of societal security and future prospects.

One such exemplar is in Alwar District, to the north east of Rajasthan. The landscape of Alwar District is semi-arid with a craggy landscape shaped by the undulating Aravalli Hills. It is here that the NGO Tarun Bharat Sangh (TBS) has been a key agent in the development of a global

Fig. 3.1 A mature johad intercepting monsoon run-off from a dry hill slope, contributing to moisture in a formerly treeless and unproductive valley bottom near Harmeerpur village. (Image © Dr Mark Everard)

exemplar programme of community-based catchment regeneration.[6,7] The founder of TBS, Rajendra Singh, began working with local people in 1985 against a backdrop of economic and ecological decline and rural depopulation. Increasing use of mechanised water extraction techniques and abandonment of traditional collaboration to enhance the recharge of groundwater during limited periods of monsoon rainfall led to an aridifying landscape and loss of perennial flows in rivers. The initial focus of TBS was on education. However, this changed when a village elder told Rajendra Singh that the fundamental issue was lack of water, not of education. Singh and colleagues took advice from a lower-caste older lady to restore or create small, localised traditional WHSs known as 'johadi' (plural of 'johad') to intercept run-off during monsoon rains, allowing it to percolate into and recharge groundwater (see Fig. 3.1).

The first TBS-initiated johad was a small structure hand-dug in 1987 in collaboration with villagers of Gopalpura. Though outcomes were uncertain, this first johad functioned as hoped, restoring soil moisture and ecology for improved food production, rejuvenating local grazing and

[6] Sinha, J., Sinha, M.K. and Adapa, U.R. (2013). *Flow—River Rejuvenation in India: Impact of Tarun Bharat Sangh's Work*. SIDA Decentralised Evaluation 2013:28. Swedish International Development Cooperation Agency, Stockholm.

[7] Everard, M. (2015). Community-based groundwater and ecosystem restoration in semi-arid north Rajasthan (1): Socio-economic progress and lessons for groundwater-dependent areas. *Ecosystem Services*, 16, pp. 125–135.

Fig. 3.2 A check dam in Kumbhalgarh Wildlife Sanctuary, Rajasthan, built to arrest monsoon flows in an ephemeral stream, enabling groundwater percolation. (Image © Dr Mark Everard)

other vegetation, and re-establishing some vitality to the Sarsa River.[8] Interest in constructing WHSs followed from adjacent parched, depopulating villages. Early TBS efforts addressed natural resource conservation methods through local community participation, but activities expanded progressively as it attracted funds, primarily from international donors, resulting in demand-led construction of hundreds of johadi based on the Gandhian ethos of *Jal Swaraj* ('water self-governance').[9]

Other context-specific WHSs have also been devised and implemented, innovating on the basis of traditional solutions. Anicuts, comprising flat bunds built to attenuate water flows across low-topography valleys, can retain bodies of surface water during monsoon rains that recharge groundwater and moisten and carry nutrients into soils subsequently cropped for mustard, chana (chick peas), bindi (okra or lady's fingers) and wheat. Check dams intercept episodic monsoon flows in steeper monsoon rivers, promoting percolation into groundwater (Fig. 3.2). WHSs consequently vary widely in design and size. They always serve the primary purpose of groundwater recharge, but many

[8] Singh, R. (2009). Community driven approach for artificial recharge—TBS experience. *Bhu-Jal News Quarterly Journal*, 24(4), pp. 53–56.

[9] Jayanti, G. (2009). *25 Years of Evolution: Restoring Life and Hope to a Barren Land*. Tarun Bharat Sangh, Alwar.

store additional surface water for livestock watering and other uses throughout the dry season. Tree-planting and regeneration on degraded hill slopes was also undertaken to restore catchment hydrology, some trees also shading johadi to reduce evaporation.

TBS co-design and support for villagers in the construction and management of WHS schemes is always demand-driven and tuned to local landscapes, community needs, traditional knowledge and available budgets. WHSs consequently vary widely in design and size. Typically, TBS contributes 30–70% of the costs of a WHS, with villages contributing not only financially but also in the form of *shramdan* ('sweat equity': collective labour for common good linked to Gandhian ideals of self-sufficiency).

The initial focal area for TBS activities was in the Arvari, Sarsa and Baghani basins (Fig. 3.3), reinstituting village governance arrangements. The rebuilding of community-based social capital constitutes a central success factor in groundwater management here, as also observed elsewhere across the world.[10] Construction and management of WHSs resurrects

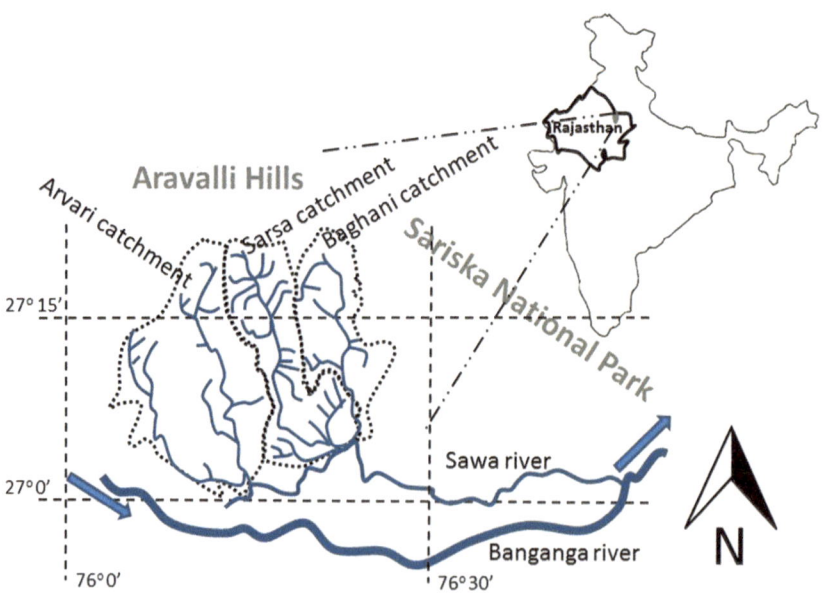

Fig. 3.3   Location of the Arvari, Sarsa and Baghani catchments in north Rajasthan. (Image © Mark Everard)

traditional knowledge and technologies and the social infrastructure necessary to operate them. The importance of social infrastructure cannot be overstated, the process of building and maintaining johadi is dependent upon resurrection of traditional village institutions.[11] Significant amongst these traditional village decision-making bodies are *Gram Sabha* ('village councils').[12] Whilst some Gram Sabha became dormant after johad construction, many remain active and progress to tackle related issues including protecting forests, building schools and other developmental works,[13] as well as zoning and regulating land uses to avoid ecological and socioeconomic degradation. At village scale, the distribution of benefits and the shares of costs of WHS construction and management are a key issue. This includes agreements about the zoning of grazing on common lands and the proportions of investment in WHSs required from those most directly benefitting from cropped lands and wells.

However, responding solely to the demand from villages local to water sources potentially risks fragmenting action across landscapes.[14] From 1998, TBS initiated a more integrated approach, forming an *Arwari Pad Yatra* ('Arvari Water Parliament') to determine water sharing and management issues across the Arvari catchment. The responsibilities of this parliament also included dispute resolution and activities such as reforestation.[15,16] Further catchment-scale arrangements between villages in progressively

[10] Lopez-Gunn, E. (2012). Groundwater governance and social capital. *Geoforum*, 43(6), pp. 1140–1151.

[11] Kumar, P. and Kandpal, B.M. (2003). *Project on Reviving and Constructing Small Water Harvesting Systems in Rajasthan.* SIDA Evaluation 03/40. Swedish International Development Cooperation Agency, Stockholm.

[12] Jayanti, G. (2009). *25 Years of Evolution: Restoring Life and Hope to a Barren Land.* Tarun Bharat Sangh, Alwar.

[13] Kumar, P. and Kandpal, B.M. (2003). *Project on Reviving and Constructing Small Water Harvesting Systems in Rajasthan.* SIDA Evaluation 03/40, Swedish International Development Cooperation Agency, Stockholm.

[14] Kumar, P. and Kandpal, B.M. (2003). *Project on Reviving and Constructing Small Water Harvesting Systems in Rajasthan.* SIDA Evaluation 03/40, Swedish International Development Cooperation Agency, Stockholm.

[15] Rathore, M.S. (2003). *Community based Management of Ground Water Resources: A Case Study of Arwari River Basin.* Institute of Development Studies, Jaipur.

[16] Jayanti, G. (2009). *25 Years of Evolution: Restoring Life and Hope to a Barren Land.* Tarun Bharat Sangh, Alwar.

regenerated catchments—the Bhagani-Teldehe, Arvari, Jahajwali, Sarsa and upper Ruparel—have restored perennial water bodies and rural livelihoods, promoting the repopulation of villages.

By 2010, responding to the specific geography and needs of each village, TBS was working with more than 700 villages in Rajasthan, with many hundreds of WHSs built and maintained by villages. Collective efforts had restored perennial water to these and other adjacent catchments by linking up village governance institutions into catchment parliaments. Singh has remarked that *"We never realised that we were recharging a river. Our effort was just to catch and allow water to percolate underground".*[17] The reappearance of perennial water bodies enabled recolonisation by aquatic wildlife, regenerating an associated set of traditional medicinal and other cultural values.[18]

Another significant success of TBS-driven community initiatives has been the empowerment of women. In 1985, women typically spent 6–7 hours daily searching for water; rising water tables and water access through hand pumps and wells close to housing now reduces this task to 5–6 minutes. Freed from the drudgery of traditional roles foraging for water, fodder and fuel, women can devote more time to tackling perceived 'social ills', contributing to health services and education (particularly of girls), and engaging in decision-making and other productive activities.[19,20] TBS has actively empowered women through enabling democratic engagement, education including Ayurvedic (traditional herbal) medicine, and formation of Women Self Help Groups (SHGs) to strengthen the role of women and to share learning across catchments.[21]

[17] Down to Earth. (1999). Coming back to life. *Down to Earth*, 15 March 1999. http://www.downtoearth.org.in/node/19493, accessed 5 March 2016.

[18] Everard, M. (2016). Community-based groundwater and ecosystem restoration in semi-arid north Rajasthan (2): Reviving cultural meaning and value. *Ecosystem Services*, 18, pp. 33–44.

[19] Kumar, P. and Kandpal, B.M. (2003). *Project on Reviving and Constructing Small Water Harvesting Systems in Rajasthan*. SIDA Evaluation 03/40. Swedish International Development Cooperation Agency, Stockholm.

[20] Jayanti, G. (2009). *25 Years of Evolution: Restoring Life and Hope to a Barren Land*. Tarun Bharat Sangh, Alwar.

[21] Rathore, M.S. (2003). *Community based Management of Ground Water Resources: A Case Study of Arwari River Basin*. Institute of Development Studies, Jaipur.

To spread the lessons of these local successes to national scale, TBS launched the *Rashtriya Jal Biradari* ('National Water Brotherhood') in 1998, comprising individuals from diverse backgrounds across India sharing concerns about water, forest and soil conservation and the re-establishment of community water rights through awareness programs. *Jal Sammelans* ('water conferences') were instigated, aimed at influencing and developing people-oriented national and state water policy.

Significant challenges were initially encountered in reconciling national and state government aspirations with effective localised solutions. State and national government perspectives on water management differed from those of TBS, *Gram Sabha* and *Pad Yatra*, even if the overarching aspiration of water self-sufficiency was shared. Under India's national legal framework, water is owned by the state, which also has sole control of its management. Community-based restoration of water management structures and institutions is therefore technically illegal without explicit consent from the state. So too is water retention for community use under the Rajasthan Drainage Act of 1956, which specifies *"Water resources standing collected either on private or public land (including groundwater) belong to the Government of Rajasthan"*. Further conflict between villages regenerating their water systems and state government revolved around the natural recolonisation of perennial water bodies by fish. As fisheries also fall legally under government control, Rajasthan state government issued a license permitting fishing rights to a water body on the Arvari catchment in 1996 to a contractor. This contract was perceived by local people as not only below market value but also as a take-over of their community resource rights, and which could ultimately lead to dispossession of community-regenerated water resources.[22,23] Villagers consequently denied access to the fish, leading to conflict with both the contractor and state government.

[22] Singh, R. (2009). Community Driven Approach for Artificial Recharge—TBS Experience. *Bhu-Jal News Quarterly Journal*, 24(4), pp. 53–56.

[23] Sinha, J., Sinha, M.K. and Adapa, U.R. (2013). *Flow—River Rejuvenation in India: Impact of Tarun Bharat Sangh's Work.* SIDA Decentralised Evaluation 2013:28. Swedish International Development Cooperation Agency, Stockholm.

This confrontation culminated in government annulling the contract. By contrast, villagers in Mandalwass retained control of their fisheries in their water management structures for community benefit, granting a licence for fishing to provide income to maintain their dam.[24] Reclamation of rights to commons from state or private agency control has become increasingly widespread amongst indigenous people across India and more widely, NGOs playing significant roles in mobilising citizens.[25,26]

TBS-promoted work has increased water availability. This has enabled diversification of cash crops and livestock composition producing significant economic gains, greater drought resilience, reduced soil erosion and distress migration, and forest regeneration aided by village-level agreements on forest exploitation and grazing.[27] Preliminary evidence of regreening of this formerly parched and treeless landscape is provided by analysis of remote sensing data.[28] An engineer's report on the outcomes of the TBS programme concluded that one of these types of WHS, the johad suited to the sloping terrain and hard underlying geology of hill lands in Alwar District, *"...are, by and large, engineering-wise sound and appropriate"*, and that *"...there can be no better rural investment that on Johads"*.[29]

Notwithstanding these former conflicts, restoration of community-based management building on traditional techniques demonstrably makes contributions to state and national goals relating to sustainable

[24] Everard, M. (2015). Community-based groundwater and ecosystem restoration in semi-arid north Rajasthan (1): Socio-economic progress and lessons for groundwater-dependent areas. *Ecosystem Services*, 16, pp. 125–135.

[25] Fenelon, J.V. (2012). Indigenous peoples, globalization and autonomy in world-system analysis. In: Babones, S.J. and Chase-Dunn, C. (Eds.), *Routledge Handbook of World-Systems Analysis*. Routledge, New York, pp. 304–312.

[26] Subramaniam, M. (2014). Neoliberalism and water rights: The case of India. *Current Sociology*, 64(2), pp. 393–411.

[27] Rathore, M.S. (2003). *Community based Management of Ground Water Resources: A Case Study of Arwari River Basin*. Institute of Development Studies, Jaipur.

[28] Davies, T., Everard, M. and Horswell, M. (2016). Community-based groundwater and ecosystem restoration in semi-arid north Rajasthan (3): Evidence from remote sensing. *Ecosystem Services*, 21(A), pp. 20–30.

[29] Agrawal, G.D. (1996). *An Engineer's Evaluation of Water Conservation Efforts of Tarun Bharat Sangh in 36 villages of Alwar District*. Tarun Bharat Sangh, Alwar.

land and water systems, drought resilience, erosion control, regeneration of forests, wildlife and the socio-economic status of villages. Today, lessons learned from community-based restoration of water systems across Rajasthan, often supported by NGOs, are informing the state government's substantial *Mukhya Mantri Jal Swavlamban Abhiyan*[30] (MJSA: 'Chief minister's water self-sufficiency mission') programme promoting water harvesting and management solutions at village scale. By co-learning, there is increasing synergy in the quest for sustainable solutions, informing continuing policy reform and refinement to increase coherence between government aspirations and effective localised solutions.

Rajendra Singh is often referred to in the media as the 'Waterman' or 'Rivermaker'. He was given the Asia-wide Ramon Magsaysay Award for Community Leadership[31] in 2001 for his community leadership work,[32] and the prestigious global 2015 Stockholm Water Prize.[33] Further scheme recognition includes the congratulation of local communities in 2000 by then Indian President KR Narayanan for minimising drought consequences and averting distress migration witnessed elsewhere in rural India, and the return of farmers and their families to villages they had previously abandoned.[34,35]

3.1.2 Square-by-Square Rehydration of Saline Drylands in Rajasthan

A solution to water challenges in one region, even within the same state, cannot safely be transferred to a different geographical context. Some

---

[30] http://mjsa.water.rajasthan.gov.in/, accessed 8 January 2019.

[31] www.rmaf.org.ph.

[32] Ramon Magsaysay Award Foundation. www.rmaf.org.ph, accessed 6 March 2016.

[33] SIWI. (2015). *The Water Man of India Wins 2015 Stockholm Water Prize*. Stockholm International Water Institute. http://www.siwi.org/prizes/stockholmwaterprize/laureates/2015-2/, accessed 6 March 2016.

[34] The Hindu. (2000). Hope in the midst of loss. *The Hindu*, 25 June 2000. http://www.thehindu.com/thehindu/2000/06/25/stories/1325041a.htm, accessed 1 October 2015.

[35] Kumar, P. and Kandpal, B.M. (2003). *Project on Reviving and Constructing Small Water Harvesting Systems in Rajasthan*. SIDA Evaluation 03/40. Swedish International Development Cooperation Agency, Stockholm.

80 km to the west of the city of Jaipur, and 180 km south-west of the TBS ashram at Bheekampura, lies the village and vicinity of Laporiya. Laporiya and its surrounding land spans an overall area of approximately 51 km² in the Jaipur District of Rajasthan, with a population of 1,764 in 236 households.[36] As with surrounding villages, research has shown that groundwater had receded to as much as 50 metres (152 feet),[37] and Laporiya had formerly also suffered a familiar pattern of social, economic and environmental decline in the 1970s and 1980s after the introduction of energised pumps, depleting usable water resources.

In contrast to the hill country in which johadi, anicuts and check dams had proven so effective in intercepting run-off down nullah (drainage lines), the landscape of Laporiya is flat and rural with slopes of only 3–4%. It is also situated in the salt lake region of Rajasthan, where saline groundwater lies close to the surface. New water management solutions had therefore to be innovated to address the local context. This was a challenge grasped by the local, Laporiya-based NGO *Gram Vikas Navyuvak Mandal, Laporiya* (GVNML: 'village growth youth board, Laporiya'). Local man Lakshman Singh established several volunteer-based groups in the early 1970s, mainly as youth clubs to repair roads, dig wells and undertake social programmes for the improvement of the village and its surroundings. A key early intervention was the manual repair of a bund to retain monsoon run-off. This work was initially undertaken solely by Singh but, to his surprise, his efforts rapidly attracted the support and volunteer labour of many more villagers. This eventually led to formation of GVNML in 1986, driven primarily by a deep-seated passion for nature.

Without formal education in water management, Lakshman Singh and colleagues pursued local knowledge about the management of moisture through small-scale water harvesting. GVNML's early experiments with ponds were unsuccessful as the deeper water drowned grasses and

---

[36] 2011 Census of India. www.censusindia.gov.in/.
[37] Sharma, S. (2016). Laporiya: A village which is saving raindrops for thirty years. *The Economic Times*, 1 May 2016. http://economictimes.indiatimes.com/news/politics-and-nation/laporiya-a-village-which-is-saving-raindrops-for-thirty-years/articleshow/52061701.cms, accessed 19 June 2017.

insects, and the intense monsoon rains washed away bunds. The locally effective solution eventually innovated by GVNML was the 'chauka' system.[38] Chauka are constructed on communal, non-arable areas on relatively permeable sandy loam and loam soils with a slope no greater than 0.5–2% on problematic saline (salty) and sodic (containing a high concentration of sodium) soils, retaining moisture to regenerate mixed grass and fodder tree pasture. Chauka comprise interconnected matrices of low bunds on three sides of a rectangle ('chauka' means 'four corners'), constructed using soil from shallow borrow pits. These bunds break long slopes into short, interconnected cells, reducing the velocity and erosive force of monsoon run-off and slowing the water sufficiently to promote infiltration into the soil and underlying groundwater.[39] Chauka bunds are no more than 60 centimetres (2 feet) high, typically extending 40–60 m across the slope with side bunds of 25–40 m (Fig. 3.4). Small size and low height are essential to avert too much water pressure eroding the bunds, with spillways at their edges allowing the free flow of water from field-to-field. Borrow pits from which earth is dug to build the bunds are typically 3 × 1.6 × 0.3 m in dimension (Fig. 3.5). Surplus run-off downstream of networks of chauka are typically diverted into naadi (storage ponds).

Water depth in chauka is never excessive, otherwise soil animals and grass roots would be drowned. Typically, maximum depth when full is only a quarter of a metre (9 inches) above ground though, as borrow pits may be up to 30 centimetres (12 inches) deep, their bases may be as much as half a metre (21 inches) below maximum monsoon water level.[40] Not only is groundwater recharged with fresh monsoon run-off, but the soils within the chauka are enriched supporting improved communal grazing. To promote this process, cow dung is left in situ to rebuild organic and nutrient content, nourishing the soil and its microorganisms. Also, by Village Development Committee (VDC) consensus,

[38] Everard, M. (2018). *Regenerative Landscapes: Rejuvenation of Linked Livelihoods and Catchment Ecosystem Services.* RICS Research, London. https://www.rics.org/uk/news-insight/research/research-reports/regenerative-landscapes/.

[39] Mahnot, S.C., Singh, P.K. and Chaplot, P.C. (2012). Chauka system—An innovative technique. In: Mahnot, S.C., Singh, P.K. and Chaplot, P.C. (Eds.), *Soil and Water Conservation and Watershed Management.* Apex Publishing House, Udaipur/Jaipur. Chap. 12.9, pp. 217–224.

[40] Padre, S. (2008). The magic of Lapodiya's chowka. *Civil Society*, 5(5), pp. 12–13.

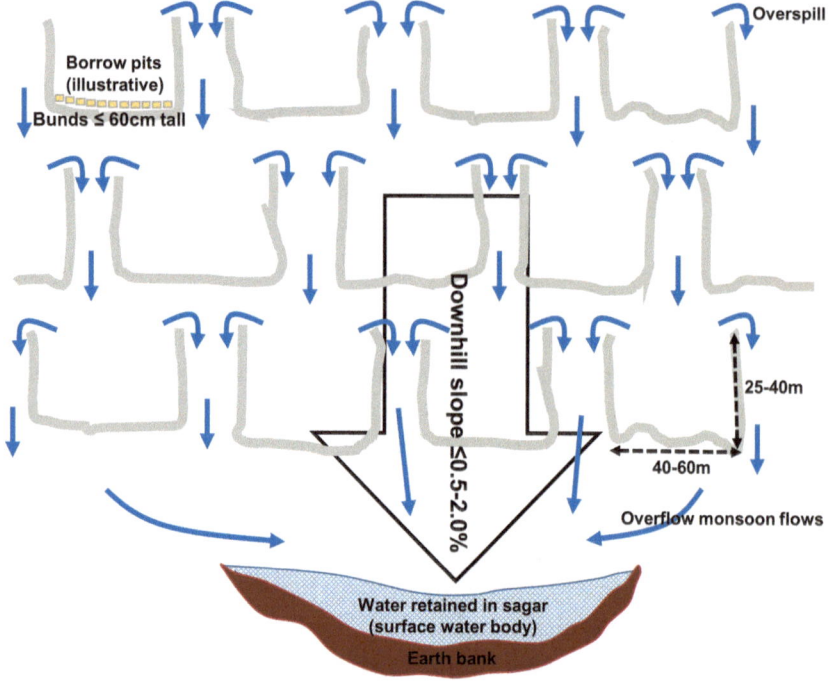

Fig. 3.4 Schematic of chauka construction and surface run-off flows. (Image © Mark Everard)

no grazing of chaukas is permitted for the month-and-a-half following the monsoon, enabling the grass to flourish and produce seed. There is also a consensual ban on allowing goats to graze on the beans of desi babool (*Vachellia nilotica*) trees, or collection of beans dropped from the tree, until decided by the VDC, again to promote ecosystem regeneration. These consensual agreements developed by grazers are aimed at maximising the benefits of chauka on the communal pasture around Laporiya, and other villages where the system is now being implemented. Some chaukas have deeper wells dug within or adjacent to them, particularly in drier, up-slope areas of the rectangle, promoting percolation of water through the soil and replenishment of the water table. Typically, communal grazing lands occupy 18–20% of village area in this region.

Fig. 3.5 A freshly dug chauka showing the low bund constructed from soil removed a shallow series of borrow pits cells, the whole chauka slowing and retaining monsoon run-off to increase soil moisture and percolation into aquifers. (Image © GVNML)

In some water recharge basins, particularly larger sagar, flood-retreating crops are grown on productive soil exposed as water levels decline during the drier months. (Flood-retreating cropping is a common production method in dry regions with seasonally variable rainfall that exploits stored soil moisture in the margins of water bodies for cropping as water levels recede.)

Chauka put control of water into the hands of villages with everyone participating in decision-making, rather than fostering a reliance on government and on transfers from dam schemes.[41] A chauka system typically has a ten-year maintenance cycle of bund repair and desilting of pits, with an additional annual desilting and repair cycle agreed at *Pad Yatra* (village gatherings) to which every villager contributes by *shramdan* (honourable physical labour looking for no reward). Chauka management, as many development initiatives in the village, are managed consensually through a VDC that, though not recognised by the state, governs all decisions pertaining to natural resources in Laporiya. The state does recognise

---

[41] Baruah, T. (2008). Water warrior. *Harmony: Celebrating Age*, pp. 56–59.

the *Gram Panchayat* (the community governance structure) for Gagardu, covering the four villages of Laporiya, Sinodiya, Gaiga and Gagardu. However, although the *Gram Panchayat* is the route by which government allocates development funds, Laporiya village resists excessive Gram Panchayat control as the Panchayat is seen as implementing government policies and making money, for example tending to auction rights rather than primarily allocating resources to villagers. This mistrust of higher tiers of formal government also extends to refusal to designate forest areas in the village, ensuring that they do not fall under the control of the state Forest Department. Physical works on water-harvesting structures (WHSs) in Laporiya were largely competed by the late-2000s, but substantial work still continues in developing institutions, gender equity, governance systems and further social progress.

Net outcomes of chauka development and other linked water management structures and institutions include drought-proofing the rain-starved village of Laporiya.[42] The slowing of flows also reduces flood risk, and the loss of soil quality and quantity through erosion. The influence of GVNML is undoubtedly enhanced by its leaders coming from a Thakur family (formerly part of a ruling dynasty, though this influence is eroding substantially in contemporary India). However, outcomes from chauka and other water management interventions for water security and the regeneration of wildlife and livelihoods has proven a more powerful driver of acceptance and wider dissemination. In 2008, the NGO WaterHarvest (then known as Wells for India) started funding the extension of chauka development into other villages in the region, as a proven means to reverse cycles of drought, hunger, caste violence and migration to cities in this particular set of geographical conditions.[43] GVNML has been successful in attracting additional international aid investment, although in the early stages of the project chauka implementation was funded solely by villagers. At present, roughly 75% of investment in chauka construction still remains through voluntary village labour. Chauka have now been tested on hundreds of hectares of land in 58

[42] Anand, R. and Anand, U. (2017). Wizard with water. *Civil Society*, 4(11–12), pp. 14–17.
[43] Baruah, T. (2008). Water warrior. *Harmony: Celebrating Age*, pp. 56–59.

mainly neighbouring villages.[44] Compared to the water and food insecurity of many adjacent villages, 75% of families in Laporiya have a marketable surplus of milk that is used, amongst other purposes, to buy fodder.[45] A participatory approach that seeks to provide for local needs is an important aspect of chauka design. GVNML has produced a *Chauka manual*[46] to promote the approach.

As with other effective WHSs, water management solutions are planned according to the workings of the water cycle and implemented communally. Effective collaboration is vital as groundwater even a few metres below the soil surface in this region is highly saline. The replenishment of shallow groundwater is therefore critical for avoiding groundwater contamination, with many areas around adjacent villages visibly lacking vegetation owning to saline soil conditions. Nevertheless, despite all of these interventions, some of the 103 wells in Laporiya have high fluoride concentration due to the underlying geology, but these are known and used only for agriculture and not for drinking.

In Laporiya, chauka have proved critical in providing for the livestock watering and grazing needs of local communities, recharging wells for year-round access and also preventing underlying saline groundwater rising to the surface and contaminating these uses. As a dryland region, most of the arable cropping is in the khariff (wet, post-monsoon) season, but there are significant areas particularly downstream of sagar (surface water bodies) where replenished groundwater supports spray irrigation for a rabi (dry season) crop (typically chilli, tomato, aubergine, onion and wheat). In some areas, there is also a third jayad (summer) crop. Private owners of land in better-watered land downstream of naadi can produce from 250 to 1,000 kg per hectare of rabi crop, depending on the rainfall in any specific year. By contrast, most adjacent villages lacking chauka systems tend only to produce a khariff crop as the land is too arid in the long drier season.

---

[44] Mahnot, S.C., Singh, P.K. and Chaplot, P.C. (2012). Water erosion control measures on non-arable lands. In: *Soil and Water Conservation and Watershed Management.* Apex Publishing House, Udaipur/Jaipur. Chap. 21, pp. 205–238.

[45] Padre, S. (2008). The magic of Lapodiya's chowka. *Civil Society*, 5(5), pp. 12–13.

[46] GVNML. (n.d.). *The Chauka Manual.* Gram Vikas Navyuvak Mandal, Laporiya (GVNML), Laporia, Rajasthan.

The spiritual importance of water is in evidence throughout the village. For example, shrines to a local matriarch and other goddesses are found by some wells. Devoothni gyaras (the 11th day after Diwali) is considered very holy, with people in Laporiya paying particular regard to the symbolism of water. Another innovation and goal of GVNML is to identify 10–15% of village and private land that can be reserved purely for wildlife as enclosed 'ecoparks'. Two 'ecoparks' have been fenced off in Laporiya, from which people or livestock are excluded, with the rapidly revegetating land reserved entirely for nature. The 'ecoparks' provide numerous co-benefits including providing seed banks, regenerating bird populations promoting seed dispersal (particularly neem tree seeds spread by parrots), hosting pollinators and pest predators, improving landscape porosity thus enhancing groundwater recharge, and they are also imbued with spiritual importance. Villager concerns about the need to remove household pests but a desire not to kill them also led to the establishment an additional nature reserve area known as a chuha gher ('mouse house'), where rodents trapped in households are taken for release into the wild. Other GVNML-promoted initiatives include roof water harvesting, and the construction of flat check dams on common land to promote water infiltration into soil.

Local testimonies indicate the impact of GVNML work: Lakshman Singh and GVNML colleagues report that there were no trees in and around Laporiya thirty years ago, while in the present day the landscape is extensively tree-covered and there is an abundance of birds. Chauka systems on common land have also increased fodder trees and grasses by at least five times over ten years.[47] This has enabled the diversification of crops from traditional dryland species such as chana (chick peas) and dahl (lentils) to crops that are heavily water-dependant such as rice, potatoes and wheat. Enhanced livestock health and yield and the quality of milk have improved local health and income resulting from measures such as a cumulative organic fertilisation of 325 ha and grass seeding on 1,600 ha between 1978 and 2009. In addition, a range of health programmes has been promoted, such as midwife training, vaccination, food distribution, and women's and children's health.[48]

---

[47] Wells for India. (2016). Before and after. *WaterWise*, 62(Autumn 2016), p. 10.

[48] GVNML. (2009). *GVNML Annual Report 2008–2009*. Gram Vikas Navyuvak Mandal, Laporiya (GVNML), Laporia, Rajasthan.

Laporiya village is an example of environmental governance that combines traditional and religious practices with scientific concepts addressing ecosystem regeneration as a foundation for tackling the long-running social and economic challenges associated with climate change.[49] Unlike many villages in rural Rajasthan reliant upon government-supplied water tankers in summer, levels of fresh groundwater have recovered from inaccessible depths to 4.5–12 metres (15–40 feet) beneath the ground surface, providing a water surplus enabling Laporiya to supply water to 10–15 surrounding villages.[50] The chauka approach, though a relatively recent innovation to deal with specific local geography, builds on lessons from land management practices that have been used for centuries. The approach can improve local resilience to drought through an integrated water resources management approach that is supported by appropriately reformed policies and investment.[51] The successes of the chauka system in dealing with the particular stresses associated with the flat topography and shallow saline groundwater have attracted the attention of the Rajasthan government.

3.1.3 Solutions and Leadership for Water Security in Rajasthan

The solutions in Alwar District and Laporiya, now spreading more broadly in influence, have been significantly driven by visionary and pragmatic NGOs. The NGO WaterHarvest,[52] operating for much of its history since founding in 1987 as 'Wells for India', has worked on a variety of community-based water projects. These have supported TBS and

---

[49] Mathur, S. (2014). A village adapts to climate change in myriad ways. *India Climate Dialogue.* http://indiaclimatedialogue.net/2014/10/30/village-adapts-climate-change-myriad-ways/, accessed 5 June 2016.

[50] Sharma, S. (2016). Laporiya: A village which is saving raindrops for thirty years. *The Economic Times,* 1 May 2016. http://economictimes.indiatimes.com/news/politics-and-nation/laporiya-a-village-which-is-saving-raindrops-for-thirty-years/articleshow/52061701.cms, accessed 19 June 2017.

[51] Narain, P., Khan, M.A. and Singh, G. (2005). *Potential for Water Conservation and Harvesting against Drought in Rajasthan, India.* Working Paper 104 (Drought Series: Paper 7). International Water Management Institute (IWMI), Colombo, Sri Lanka.

[52] https://www.water-harvest.org/.

GVNML as well as many other local, community-facing NGOs active across Rajasthan and into the adjacent Indian state of Gujarat. Many of the communities supported by WaterHarvest are in the Thar Desert, where people have adapted water harvesting to the extremely dry and changing climate, and a growing human population, throughout a period of centuries.[53] WaterHarvest recognises ten agroclimatic zones in Rajasthan alone, and has collaborated on the guide book *Wise Water Solutions in Rajasthan*[54] (produced in English, Hindi and Gujarati) documenting a diversity of water management solutions founded on traditional technologies, or modern innovations based on underpinning traditional knowledge. *Wise Water Solutions in Rajasthan* addresses the appropriateness of each type of solution to specific contexts, with basic information about construction and operational considerations. Some examples of these techniques are outlined in Box 3.1.

Box 3.1 Flexible Water Management Solutions Adapted to Rajasthan's Varying Geography and Cultures

The following are a subset of traditional and innovative techniques reviewed in the guide book *Wise Water Solutions in Rajasthan*.[55]

• Taanka are innovations for conditions in the Thar Desert of central and western Rajasthan, where rain is sporadic and sparse and the underlying sand is unable to retain water. WaterHarvest has promoted, and now widely supports, the construction and use of taanka. Taanka are a household and community-scale water harvesting and safe storage structure comprising a concrete-sided and covered well. This well is recharged from monsoon rainfall onto a circular, concreted microcatchment depression in the ground, bounded by thorny vegetation to avoid animal incursions. This simple technology captures and stores rainfall for year-round access, meeting the needs of households at low capital cost and with minor maintenance requirements.

---

[53] Jal Bhagirathi Foundation and Wells for India. (2010). *Adapting Water Harvesting to Climate Change*. Jal Bhagirathi Foundation/Wells for India, Jodhpur/Udaipur.

[54] Sharma, O.P., Everard, M. and Pandey, D.N. (2018). *Wise Water Solutions in Rajasthan*. WaterHarvest/Water Wise Foundation, Udaipur, India.

[55] Sharma, O.P., Everard, M. and Pandey, D.N. (2018). *Wise Water Solutions in Rajasthan*. WaterHarvest/Water Wise Foundation, Udaipur, India.

- Loose stone check dams, also known as gully plugs and pagaras, are widely build in nullah (drainage lines) in areas of moderate slope with low to moderate geological permeability (Fig. 3.6). They detain both water and sediment to increase soil moisture and build soil content in upper catchments.
- Rooftop rainwater harvesting structures are independent of geology and topography but serve as vital water capture techniques for provision of household water supply. They can recharge surface or subsurface storage tanks, collecting water harvested during monsoons for year-round access.
- Step wells are known by many alternative local names, including baoli, bavadi, jhalras and open wells. Step wells allow stepped access to receding groundwater, enabling access to fluctuating groundwater levels for drinking and other uses. Some also serve spiritual needs.
- Drip irrigation systems are a modern innovation, suited to cropping on sandy and coarse loamy soil. They are highly efficient in the use of water by targeting the roofs of crops. Drip irrigation enables a much more efficient use of water compared to traditional flood irrigation methods.
- Bio Sand Filters are a low-technology, cheap and readily constructed household-scale solution for treating water to remove chemical and microbial contaminants. These structures are a simple, nature-based approach allowing water to run slowly under gravity through a column of sand, typically retained in a concrete structure, close to households where the water is used.

In experimental watersheds in India, including the Bundi watershed in Rajasthan, water levels in wells close to community-constructed and maintained WHSs improved groundwater yield both quantitatively and in terms of duration compared to more remote wells.[56] Groundwater level in the Bundi watershed rose by 5.7 m, with a corresponding 66% increase in irrigated area.[57]

---

[56] Wani, S.P., Sudi, R. and Pathak, P. (2009). Sustainable groundwater development through integrated watershed management for food security. *Bhu-Jal News Quarterly Journal*, 24(4), pp. 38–52.

[57] Wani, S.P., Pathak, P., Sreedevi, T.K., Singh, H.P. and Singh, P. (2003). Efficient management of rainwater for increased crop productivity and groundwater recharge in Asia. In: Kijne, W., Barker, R. and Molden, D. (Eds.), *Water Productivity in Agriculture: Limits and Opportunities for Improvement*. CAB International, pp. 199–215.

Fig. 3.6 A cascade of loose stone check dams retaining water and building fertile soil from monsoon run-off down a nullah (drainage line) in Morwaniya, Rajasthan. (Image © Mark Everard)

State governments have not been blind to these innovations and their successes. As previously addressed, the Government of Rajasthan is seeking to promote check dams, anicuts, chauka and other nature-based approaches as part of its *Mukhya Mantri Jal Swavlamban Abhiyan* programme, with the aim of empowering 21,000 villages to regain control of their local water supply by restoring former water management practices adapted to geographical, cultural and intensely episodic rainfall conditions. Government of Rajasthan recognises the *Wise Water Solutions in Rajasthan*[58] guide book as supporting MJSA. Maharashtra and Gujarat are other states with decentralised water governance policies in addition to their 'big technology' solutions to service urban and other intensive centres of demand.[59]

[58] Sharma, O.P., Everard, M. and Pandey, D.N. (2018). *Wise Water Solutions in Rajasthan*. WaterHarvest/Water Wise Foundation, Udaipur, India.

[59] Kulkarni, S. (2011). Women and decentralised water governance: Issues, challenges and the way forward. *Economic and Political Weekly*, 46(18), pp. 64–72. https://www.epw.in/journal/2011/18/review-womens-studies-review-issues-specials/women-and-decentralised-water, accessed 12 January 2019.

In 2012, the Government of India nationally adopted a revision of its initial 1987 National Water Policy (NWP), laying new emphasis on treating water as an economic good, preparedness for climate change, and noting that "*…special emphasis should be given towards mitigation at micro level by enhancing the capabilities of community to adopt climate resilient technological options*" implicitly favouring community-based water management solutions.[60] Though not supported by all states, the NWP marked a point of reflection on the tendency, still deeply entrenched amongst many of India's water professionals, to automatically favour 'big technology' solutions such as dam-and-transfer schemes. As one major example of not only entrenched by possibly disastrous centralised thinking and planning, India's National Water Development Agency[61] is pressing on with planning for an Indian Rivers Inter-link Programme. This Programme comprises a large-scale civil engineering project aiming, as the title suggests, to link Indian rivers by a network of reservoirs and canals as an engineered solution to reduce persistent floods in some parts and water shortages in others. Though ambitious, the scheme has many opponents, and runs entirely counter to conservation of the ecology and other aspects of the natural character and supportive capacities of the precious water resources of the nation.

Pressures are certainly mounting across India for a transformation in water management as late-Twentieth Century technological, policy and economic forces, compounded by population rise, urbanisation and climate change, foment grave looming problems. In late 2018, for example, a government report documented that India was experiencing the worst water crisis in its history, threatening millions of lives and livelihoods. Some 600 million Indians (about half the national population) were facing high to extreme water scarcity conditions from which about 200,000 die every year, with a prognosis that, by 2030, the country's demand for water is likely to be twice the available supply.[62]

---

[60] Government of India. (2012). *National Water Policy (2012)*. Ministry of Water Resources, Government of India. http://mowr.gov.in/sites/default/files/NWP2012Eng6495132651_1.pdf, accessed 12 January 2019.

[61] National Water Development Agency, Ministry of Water Resources, Government of India.

[62] NITI. (2018). *Composite Water Management Index: A Tool for Water Management.* National Institution for Transforming India (NITI Aayog), New Delhi. http://www.niti.gov.in/writereaddata/files/document_publication/2018-05-18-Water-index-Report_vS6B.pdf, accessed 12 January 2019.

3.1.4   Water Recharge Solutions Across Asia

Whilst the focus of the early part of this chapter has been on successes in Rajasthan, there is a rich tradition of water management solutions across Asia as well as similar, more recent problems associated with modern shifts in technology, policy and economics.

The western Indian state of Gujarat exemplifies the challenges faced by a long-established agrarian economy making a transition to energised groundwater overexploitation since the 1980s. Perverse energy subsidies have led to a similar saga of declining groundwater levels and quality, with consequent rising energy costs and carbon footprint. A community-based groundwater recharge initiative, emerging as a social movement in the Saurashtru region of Gujarat (Box 3.2), builds upon India's diversity of traditional approaches to community-based water management adapted over centuries to local geography and culture.

> **Box 3.2 Community Groundwater Recharge Initiative, Saurashtru Region, Gujarat State**
>
> Groundwater depletion in Gujarat's arid environment has serious economic and ecological consequences. A groundwater recharge initiative established in the early 1990s in the Saurashtru region has grown into a large social movement.[63]
>
> Whereas state schemes focused largely on exploitation have largely failed, successful Gandhian NGO-led community action working in concert with local institutions has focused on groundwater recharge as a basis for *jal swaraj* (water self-sufficiency). The movement spread by village-to-village contagion, with demonstrable benefits promoted by audio and video communication and as influential sectors and social elites joining the movement.

The Indian state of Maharashtra has instigated its own Jalyukt Shivar Abhiyan (JSA) programme (Box 3.3), using a smartphone app to link traditional knowledge about water management in the landscape with remotely sensed data.

---

[63] Shah, T. (2000). Mobilising social energy against environmental challenge: Understanding the groundwater recharge movement in Western India. *Natural Resources Forum*, 24, pp. 197–210.

> **Box 3.3 The Jalyukt Shivar Abhiyan (JSA) Programme in Maharashtra**
>
> Jalyukt Shivar Abhiyan (JSA) is a flagship program of Government of Maharashtra, launched in 2014 to provide long-term and sustainable solutions to the water scarcity problem faced by rural communities.[64] There is a government aim to make Maharashtra a drought-free state by 2019.
>
> The JSA programme involves deepening and widening of streams, construction of cement and earthen stop dams, work on nullahs (drainage lines) and digging of farm ponds. A mobile app, developed by MRSAC (Maharashtra Remote Sensing Application Centre), is being used to map appropriate locations. Mapped locations can be monitored through a web page. Users are able to download the application to their smartphones, and to view an instruction manual and mapping locations along with photographs. Statistics broken down by District, Taluka (also known as a Tehsi or Taluq: an administrative division below District) and by project are also available both in tabular and graphic form. The JSA project aims to make 5000 villages free of water scarcity every year.

Farmers in Pakistan's Punjab province have also succeeded in regreening their lands and developing resilience against increasingly extreme weather conditions and erratic rainfall through rainwater harvesting using small dams, restoring water to the landscape and hope to farmers[65] (Box 3.4). This is particularly significant for the region as agriculture is the largest employment sector in Pakistan, involving 44.7% of total national manpower and contributing 23% to national GDP. Beyond trading considerations, agriculture is the mainstay of subsistence in rural Pakistan. Yet with as much as 70% of annual rainfall occurring only in the monsoon season, there is a long history of groundwater depletion, particularly for the 60% of the population living in rural and hilly areas where it is necessary to manage water at a highly localised scale.[66] Whilst

---

[64] http://mrsac.maharashtra.gov.in/jalyukt/.

[65] Pakistan Water Partnership. (2016). Farmers in Pakistan's Punjab province are greening their lands and combating weather vagaries through rainwater harvesting using small dams. *Pakistan Water Partnership*, 4 January 2016. http://pwp.org.pk/?p=788, accessed 6 March 2016.

[66] Ghani, M.W., Arshad, M., Shabbir, A., Shakoor, A., Mehmood, N. and Ahmad, I. (2013). Investigation of potential water harvesting sites at Potohar using modelling approach. *Pakistani Journal of Agricultural Science*, 50(4), pp. 723–729.

Box 3.4 Rehydration of the Pakistani Punjab Through Small
Dams[67]

Small dams to harvest rainwater are being built by an increasing number
of farmers across the Pakistani Punjab, providing water supplies supporting improved yields and reliability of cropping and water for livestock
throughout the year. They thereby also reduce dependence on other
water sources.

Rainwater harvesting raised the groundwater table from 450 feet to 200
feet in one village. Successes in pioneering villages inspired people in
nearby villages to pool money for building mini-dams to reap the benefits
of modern agriculture. Typically, a mini-dam is constructed across a natural
stream or nullah near farmland, usually with some form of pumping mechanism to pipe water across the farm. The International Fund for Agricultural
Development[68] (IFAD) has supported some initiatives, matching funding
raised by villagers.

adverse impacts of climate change are projected to exacerbate the water
crisis in Pakistan, such novel, localised and geographically appropriate
solutions offer hope of resilience and continued prosperity.

These types of community empowerment initiatives, bringing together
institutions from scientific, non-government, government and farming
sectors, have restored groundwater levels and livelihoods more widely
across Asia, reversing former cycles of linked socio-ecological degradation. Significant groundwater rises are, for example, reported where
community-based participatory methods have been developed at
benchmark sites in several Indian states/provinces, as well as in Thailand,
Vietnam and China.[69] These community-based empowerment initiatives
have restored groundwater levels, improved productivity by up to 250%,

---

[67] Pakistan Water Partnership. (2016). Farmers in Pakistan's Punjab province are greening their lands and combating weather vagaries through rainwater harvesting using small dams. *Pakistan Water Partnership*, 4 January 2016. http://pwp.org.pk/?p=788, accessed 6 March 2016.

[68] IFAD. (n.d.). *IFAD: Investing in Rural People*. International Fund for Agricultural Development (IFAD), United Nations. https://www.ifad.org/, accessed 28 May 2016.

[69] Wani, S.P., Sudi, R. and Pathak, P. (2009). Sustainable groundwater development through inte-grated watershed management for food security. *Bhu-Jal News Quarterly Journal*, 24(4), pp. 38–52.

reversed degradation of natural resources, and substantially improved the livelihoods of poor people in 368 experimental watersheds across Asia.[70,71]

3.1.5 Other Recharge Initiatives Across the Developing World

A striking African example of landscape resuscitation through localised water management techniques is seen in the regeneration of farmland in a dry region of south-central Zimbabwe. Local man Zephaniah Phiri Maseko from Zvishavane has become known and widely consulted globally for innovations, implementation and sharing of effective localised water management techniques in indigenous permaculture and drought-sensitive farming methods.[72] These efforts have seen him often referred to as 'The Water Harvester' or 'The man who farms water'.[73] Not unlike Rajendra Singh in Rajasthan, Phiri resisted state authority, spending time in gaol for flouting the law regarding who controls water management notwithstanding attaining clearly significant local beneficial outcomes.

Phiri's efforts centred on his eight-acre landholding, a formerly arid landscape comprising fragile soils with erratic rainfall. Through innovation and application of unorthodox and labour-intensive water harvesting techniques, Phiri progressively transformed it into fertile farmland and a perennial wetland. The principle method deployed is construction of 'Phiri Pits' dug along contour ridges to capture runoff directed by low

---

[70] Wani, S.P. and Ramakrishna, Y.S. (2005). Sustainable management of rainwater through integrated watershed approach for improved livelihoods. In: Sharma, B.R., Samra, J.S., Scot, C.A. and Wani, S.P. (Eds.), *Watershed Management Challenges: Improved Productivity, Resources and Livelihoods*. IMMI, Sri Lanka, pp. 39–60.

[71] Wani, S.P., Ramakrishna, Y.S., Sreedevi, T.K., Long, T.D., Wangkahart, T., Shiferaw, B., Pathak, P. and Keshava Rao, A.V.R. (2006). Issues, concept, approaches practices in the integrated watershed management: Experience and lessons from Asia. In: *Integrated Management of Watershed for Agricultural Diversification and Sustainable Livelihoods in Eastern and Central Africa: Lessons and Experiences from Semi-Arid South Asia*. Proceedings of the International Workshop held during 6–7 December 2004, Nairobi, Kenya, pp. 17–36.

[72] Witoshynshky, M. (2000). *The Water Harvester*. Weaver Press, Harare.

[73] Lancaster, B. (2008). CASE STUDY: Drought resistant farming in Africa. *The Ecologist*, 21 November 2008. http://www.theecologist.org/campaigning/food_and_gardening/360257/case_study_drought_resistant_farming_in_africa.html, accessed 7 March 2016.

bunds during erratic rains. These raised the water table, providing constant moisture for various trees, crops and ponds containing fish. Since 1987, Phiri's methods have been widely promoted through the Zvishavane Water Project, one of Zimbabwe's first NGOs, with visitors from all over the world learning from localised successes.

Methods that Phiri claims were developed by trial and error have similarities with a diversity of water management techniques founded on traditional wisdom seen across Africa, Asia, the Americas and the wider drier tropical world, supporting multiple benefits for dependent communities. These diverse global methods are reviewed in Fred Pearce's book *Keepers of the Spring*,[74] my own *The Hydropolitics of Dams*,[75] Brad Lancaster's *Rainwater Harvesting for Drylands and Beyond*[76] and IMAWESA's *100 Ways to Manage Water for Smallholder Agriculture in Eastern and Southern Africa*.[77]

In east Africa, the NGO Excellent Development[78] actively promotes 'sand dams' as a method to regenerate the water table for combined ecosystem and livelihood benefits. This approach is based on similar methods and principles to johadi or check dams in Rajasthan, leading their local promotors to describe themselves as 'oasis makers'. In Kenya, sand dams represent a low-tech means to recharge shallow groundwater, and to improve the quality of water as it is filtered by the sand. The dams themselves may be made of concrete or other resilient materials but, in sandy catchments, they fill after seasonal floods with sand. This promotes water infiltration and purification as well as soil formation. They work in locations where underlying rock strata are not highly permeable, as otherwise water would infiltrate into deeper, less accessible aquifers. Sand dams are described as representing 'low-tech weapons' to tackle

---

[74] Pearce, F. (2004). *Keepers of the Spring: Reclaiming Our Water in an Age of Globalization*. Island Press, Washington, DC.

[75] Everard, M. (2013). *The Hydropolitics of Dams: Engineering or Ecosystems?* Zed Books, London.

[76] Lancaster, B. (n.d.). *Rainwater Harvesting for Drylands and Beyond*. http://www.harvestingrainwater.com/resource-appendices/volume-1/#, accessed 23 April 2015.

[77] Mati, B.M. (2007). *100 Ways to Manage Water for Smallholder Agriculture in Eastern and Southern Africa: A Compendium of Technologies and Practices*. SWMnet Working Paper 13, IMAWESA (Improved Management in Eastern & Southern Africa), Nairobi.

[78] Excellent Development. (n.d.). Pioneers of sand dams. http://www.excellentdevelopment.com/home, accessed 23 April 2015.

the effects of a changing climate where geological and other geographical conditions are suitable.[79]

Further examples of the application of nature-based solutions for water and soil conservation are found in the Central Plateau of Burkina Faso in West Africa (Box 3.5) and in the Peruvian uplands of Ayacucho in South America (Box 3.6), both of which adapt traditional water harvesting approaches hybridised with intensive water management infrastructure.

Box 3.5 Nature-based Solutions in the Central Plateau of Burkina Faso

In the Central Plateau region of Burkina Faso, West Africa, a variety of locally adapted nature-based soil and water conservation techniques are in common use. These include, for example, small planting pits known locally as *zaï* used to restore degraded land and/or increase agricultural production, a method traditionally used by small-scale farmers in the north of the Central Plateau where rainfall is low and variable.[80] Storage of surface flow in naturally low-lying areas (*bas-fonds*) during the rainy season is another traditional technique, still widely used today along with other small reservoirs (*retenues d'eau*) across the Central Plateau primarily for livestock watering, irrigation of market gardens and recharging groundwater.[81]

Large-scale promotion and implementation of these and other similar nature-based water management techniques in the north of the Central Plateau began during the 1960s and 1970s, with support from national Government and international NGOs, responding to severe droughts.[82] National government and international organizations implemented several large projects in ensuing decades, including PATECORE (Projet Aménagement des Terroirs et Conservation des Ressources), implemented in 400 villages in

---

[79] Brahic, C. (2006). Sand dams: Low-tech weapons against climate change. *New Scientist*, 16 November 2006. https://www.newscientist.com/article/dn10587-sand-dams-low-tech-weapons-against-climate-change/, accessed 19 May 2017.

[80] Kaboré, D. and Reij, C. (2004). *The Emergence and Spreading of an Improved Traditional Soil and Water Conservation Practice in Burkina Faso*. Environment and Production Technology Division Discussion Paper. International Food Policy Research Institute, Washington, DC, USA.

[81] GoBF. (2018). *Programme National du Secteur Rural 2016–2020*. Government of Burkina Faso, Ouagadougou.

[82] Kabore-Sawadogo, S., Ouattara, K., Balima, M., Ouedraogo, I., Traore, S., Savadogo, M. and Gowing, J. (2013). Burkina Faso: A cradle of farm-scale technologies. In: Critchley, W. and Gowing, J. (Eds.), *Rainwater Harvesting in Sub-Saharan Africa*. Earthscan, Oxon, UK.

Bam province[83] with further expansion under the Government's current National Adaptation Plan[84] and Intended Nationally Determined Contributions (INDC).[85]

The repair and construction of both small and large-scale dams and reservoirs forms a major part of Burkina Faso's strategy to increase access to water, not just for small-scale farmers but for many different types of water users including diversion to cities.[86]

Box 3.6 Nature-based Solutions in the High-Altitude Ayacucho Province of Peru, South America

Located in the centre-south of Peru, the province of Ayacucho has very low rainfall (annual precipitation varies from 5 mm to 1,500 mm) and is at high altitude (the city of Ayucucho nestles in a valley in high Andean mountains at 2,761 masl). The population of Ayacucho province is predominantly agricultural, though the fragility of these semi-arid upland ecosystems renders livelihoods highly susceptible to climate change. Irregular water availability in lagoons, lakes and rivers is a particular problem, particularly in rural areas of the Andean regions of the province where 62% of agriculture is rain-fed. Nature-based solutions continue to be developed by communities who are highly dependent on natural resources as an adaptation to increasing climatic variability.

Bofedales are a common nature-based water management intervention, comprising naturally occurring depressions in the flat, treeless uplands, though many have been expanded and connected by people over thousands of years to enhance vegetation and water supply. *Qochas* are another nature-based intervention, comprising small-scale water storage basins for local use, many of which were previously abandoned by people migrating to avoid the extended civil war from the 1980s.

---

[83] PATECORE (Projet Aménagement des Terroirs et Conservation des Ressources dans le Plateau Central). (2004). Les expériences du PATECORE dans le Plateau Central au Burkina Faso, 17 Années au côté des producteurs dans la lutte contre la désertification du Bam, de l'Oubritenga et du Kourwéogo. Document de Capitalisation. Ministère de l'Agriculture, de l'Hydraulique et de Ressources Halieutiques, Ougadougou.

[84] GoBF. (2015). *Burkina Faso National Climate Change Adaptation Plan (NAP)*. Government of Burkina Faso, Ouagadougou.

[85] GoBF. (2015). *Intended Nationally Determined Contribution in Burkina Faso*. Government of Burkina Faso, Ouagadougou.

[86] GoBF. (2018). *Programme National du Secteur Rural 2016–2020*. Government of Burkina Faso, Ouagadougou.

Nature-based water harvesting and storage projects promoted by state agencies with international cooperation date back to the 1980s. These include the National Watershed Management and Soil Conservation Program (PRONAMACHCS), the Natural Resources Management Project in the Southern Highlands (MARENAS), the Sustainable Water and Soils Management Project in Laderas (MASAL), and the Adaptation to Climate Change Program (PACC Peru).

Regional agriculture and drinking water for Ayacucho city depend on the Cachi Project. The Cachi Project transfers water from the Chicllarazo, Choccoro and Apacheta rivers to the Cuchoquesera Dam, which is also fed by wetland ecosystems including constructed and natural *bofedales*. This represents a mix of nature-based solutions supporting the filling of the engineered dams and water transfer schemes supplying more intensive water users.

Recognition that natural processes do not respect national borders is important for taking a fully ecosystem-based approach. In southern Africa, the adjacent nations of Malawi and Mozambique formed a joint commitment in 2019 to collaborate on transboundary conjunctive management of groundwater (Box 3.7).

Box 3.7 Transboundary Conjunctive Water Management between Malawi and Mozambique[87]

The Shire Aquifer and River Basin system is shared by the neighbouring nations of Malawi and Mozambique in Southern Africa. The basin faces many water issues, including posing flood and drought risks to the two countries as well as problems relating to the pervasion of the invasive aquatic water hyacinth (*Eichhornia crassipes*) plants. Groundwater in particular is critical for water security for the dispersed rural communities across the basin.

The two neighbouring countries are acutely aware of their mutual interdependence in the aftermath of recent floods. Following completion of a

[87] GRIPP. (2019). *SADC member states of Malawi and Mozambique united in Commitment to Transboundary Conjunctive Water Management.* Groundwater Solutions Initiative for Policy and Practices (GRIPP), 4 July 2019. http://gripp.iwmi.org/2019/07/04/sadc-member-states-of-malawi-and-mozambique-united-in-commitment-to-transboundary-conjunctive-water-management/.

joint research project aimed at better understanding the integrated hydro-geological system and linked socio-economic development issues, approaches to conjunctive water management were devised to improve overall water security and climate change adaptation. A strategic action plan (SAP) was developed on the back of the research findings to prioritise beneficial joint actions. On this basis, the two countries are drafting a Memorandum of Understanding for data sharing, facilitating decision-making in the Shire System and shaping future conjunctive water management through transboundary collaboration.

A variety of organisations, including the Advanced Centre for Water Resources Development and Management[88] (ACWADAM) and the United Nations, seek to promote wider uptake and innovation of locally adapted indigenous methods for the linked benefits of fragile, dry ecosystems and the livelihoods of their dependents. The UN Water *World Water Development Report 2018* report[89] documents a wide range of other nature-based solutions globally that use ecosystem regeneration as a basis for socio-ecological stability and opportunity.

3.1.6   Aquifer Recharge as a Water Storage Solution

Managed Aquifer Recharge (MAR) interventions are widely adopted across India, supported by communities, State and Federal Governments, although practical outcomes are not always monitored. The 2007 budget of the Government of India allocated $US450 m (₹IN 1800 crore) to convert dry dug wells into recharge structures in 100 districts.[90] Remote sensing assessment of net benefits from substantial Government of Gujarat investment in large-scale and small-scale schemes to increase water storage and recharge found a net 29% increase in water storage

---

[88] www.acwadam.org.

[89] UN Water. (2018). *World Water Development Report 2018*. UN Water. http://www.unwater.org/publications/world-water-development-report-2018/, accessed 20 February 2019.

[90] Prathapar, S., Dhar, S., Tamma Rao, G. and Maheshwari, B. (2015). Performance and impacts of managed aquifer recharge interventions for agricultural water security: A framework for evaluation. *Agricultural Water Management*, 159, pp. 165–175.

across the state, comparing conditions pre- and post-intensification of water management, with significant increases in cropped area.[91]

However, not all upscaling (replicating at larger scale to increase overall benefits) approaches are successful, as discovered when trying to upscale measures that were successful in small, distributed scale across broad landscapes (see Box 3.8).

A more diffuse approach is being encouraged around the city of Chennai in Tamil Nadu state, south India (Box 3.9). Although the 'anchor services' for the Chennai scheme are improving fresh water security and resisting saline intrusion by recharging groundwater, there is a significant prospect of achieving wider linked socio-economic and environmental benefits through a more diffuse, landscape-scale approach.

Box 3.8 Unsuccessful Attempts to Upscale Locally Effective Infiltration Processes in Gujarat

In 2008, the Government of Gujarat established an expert Task Force to develop and recommend a Managed Aquifer Recharge (MAR) strategy. Observing that farmer-driven groundwater recharge projects tended to be small and suboptimal in net impact on water tables extending beyond village boundaries, the Task Force identified de-watered aquifers as potential underground 'reservoirs' for storing water using enlarged recharge basins. However, this upscaling approach was problematic.[92]

Economic analysis found that supply-driven options identified by hydro-geologists included construction of infiltration basins that might take decades to build and further decades to reverse groundwater deficits in the absence of proactive demand management.

Aquifer recharge for water resource storage has been developed at industrial scale in the US and Europe. However, it is important to distinguish between Artificial Aquifer Recharge (AR) and Aquifer Storage and Recovery (ASR). AR enhances groundwater recharge through man-made conveyances, such as infiltration basins, though injection wells may be

[91] Chinnasamy, P. Misra, G., Shah, T., Maheshwari, B. and Prathapar, S. (2015). Evaluating the effectiveness of water infrastructures for increasing groundwater recharge and agricultural production—A case study of Gujarat, India. *Agricultural Water Management*, 158, pp. 179–188.

[92] Shah, T. (2014). Towards a managed aquifer recharge strategy for Gujarat, India: An economist's dialogue with hydro-geologists. *Journal of Hydrology*, 518, pp. 94–107.

Box 3.9 Increased Managed Aquifer Recharge Around the City of Chennai

A more diffuse approach is being encouraged around Chennai, a city of 4.7 million people (2011 Census of India) in Tamil Nadu state, south India. Over 90% of Chennai's water supply is served by reservoirs that depend on monsoon rains. City demand for water during non-monsoonal months is mostly met by groundwater extraction once the reservoirs are emptied.[93] City expansion has put increasing pressure on both the quantity and quality of available water.

Seeking a more sustainable approach, Chennai is promoting infiltration ponds and check dams in the periphery of the city to replenish local aquifers in order to "...build a credit that can be drawn on in drought"[94] and to counter saline intrusion. Rainwater harvesting from roofs of larger buildings has also been mandatory since 2001. Many institutions and stakeholder groups express high acceptance of the groundwater recharge and management approach.[95] However, the conflicting interests of some institutions and stakeholders is hindering wider implementation of aquifer recharge. For example, there is resistance from farmers required to operate and maintain infiltration structures built on their land, providing public benefits with some government support but with relatively small self-beneficial returns.

used where there are impermeable strata between the land surface and the aquifer or where land area available for infiltration is limited. By contrast, ASR is a specific type of AR practiced to augment groundwater resources for recovery for future uses, using aquifers as a subterranean reservoir. ASR is therefore likely to produce a more limited set of ecosystem service benefits.

---

[93] Sakthivel, P., Elango, L., Amirthalingam, S., Pratap, C.E., Brunner, N., Starkl, M. and Thirunavukkarasu, M. (2014). *Managed Aquifer Recharge: The Widening Gap between Law and Policy in India*. In: Proceedings of the IWA World Water Congress, Lisbon, Portugal, September 2014, pp. 21–26.

[94] Gao, L., Connor, J.D. and Dillon, P. (2014). The economics of groundwater replenishment for reliable urban water supply. *Water*, 6, pp. 1662–1670.

[95] Brunner, N., Starkl, M., Sakthivel, P., Elango, L., Amirthalingam, S., Pratap, C.E., Thirunavukkarasu, M. and Parimalarenganayaki, S. (2014). Policy preferences about managed aquifer recharge for securing sustainable water supply to Chennai city, India. *Water*, 6(12), pp. 3739–3757.

Aquifer recharge is occurring experimentally in southern England (Box 3.10) and at operational scale in the US states of California (Box 3.11) and Arizona (Box 3.12). All three of these schemes are being undertaken for utilitarian ASR purposes, with some co-benefits but no wider ecosystem replenishment with its associated wider enhancement of ecosystem services. Where multiple uses of aquifers occur and/or where it is necessary to inject or encourage percolation of fresh water to deter salt water intrusion into freshwater aquifers and to control land subsidence, as in many regions of the United States, tight regulatory guidelines are necessary for example as guided by the US Environment Protection Agency.[96]

Box 3.10 Experimental Aquifer Recharge in Southern England

English water company Thames Water is experimenting with a borehole a quarter-kilometre deep near Horton Kirby, Kent, to understand the potential for drinking water to be stored in a 'bubble' 250 metres underground.[97]

Water storage in the deeper aquifer is also intended to relieve pressure on the River Darent, a chalk-bedded river badly affected by groundwater abstraction. However, this project is essentially a utilitarian ASR exercise storing water for subsequent abstraction for public supply rather than recharging the ecosystem for multiple benefits.

Box 3.11 The Orange County, California, Groundwater Replenishment System (GWRS)

An established Groundwater Replenishment System (GWRS) in Orange County, California, has been expanded by a $US142 million programme to inject an additional 38.2 m³ of wastewater treated by reverse osmosis annually. The expanded scheme is intended to store up to 127 m³ of water to supply households containing 850,000 people.[98] In addition to

---

[96] US EPA. (n.d.). *Aquifer Recharge (AR) and Aquifer Storage & Recovery (ASR)*. US Environmental Protection Agency. http://water.epa.gov/type/groundwater/uic/aquiferrecharge.cfm, accessed 8 April 2015.

[97] WaterActive. (2015). Thames Water brings a little bit of Las Vegas to Kent. *WaterActive*, July 2015, p. 4.

[98] Force, J. (2015). Orange county expands its groundwater replenishment system. *Water*, 21, pp. 20–22.

evaporation-free storage for later reuse, injection of treated water through recharge wells also resists seawater intrusion.

Though costly, GWRS water can be as much as half the cost of imported water, with further advantages from being locally controlled and reducing pressure on regions from which water is transferred. This GWRS is again for utilitarian storage of treated water for later extraction, also resisting saline groundwater intrusion, rather than landscape-scale ecosystem regeneration.

Box 3.12 Aquifer Storage in Arizona

Since 1994, the Arizona Water Banking Authority (AWBA), the Arizona Department of Water Resources (ADWR) and the Central Arizona Water Conservation District (CAWCD) have been engaged in an innovative program to store Colorado River water in the aquifers of central and southern Arizona.[99]

The AWBA stores nearly 4 million acre-feet (nearly five million megalitres) of water underground for recovery to meet the needs of municipal and industrial sectors and priority users of the Colorado River, including state obligations under Indian water rights settlements and inter-state obligations with Nevada. This again is a utilitarian storage ASR scheme, rather than regeneration of the wider socio-ecological catchment system.

3.1.7   Connected Thinking to Address Whole-Catchment Functioning

There is no inherent critique of heavy engineered water management technologies here, as it is impossible (at least currently) to envisage a major city living entirely on its own endogenous resources for water supply and wastewater assimilation. However schemes such as large dams,

---

[99] AWBA, ADWR and CAP. (2014). *Recovery of Water Stored by the Arizona Water Banking Authority: A Joint Plan by AWBA, ADWR and CAP.* Arizona Water Banking Authority. http://www.cap-az.com/documents/departments/planning/recovery/Joint-Recovery-Plan-FINAL4-14-14.pdf, accessed 4 June 2016.

and the transfer of water out of donor catchments to service intensive industrial, irrigation and urban demands, inevitably creates a wide diversity of externalities. These include ecosystem vitality and the needs of rural constituencies, which are all too often overlooked as asymmetrically powerful economic and political constituencies appropriate available resources.[100] A new paradigm is required that does not automatically assume that engineered approaches and technically efficient mechanisation are universally beneficial, largely by overlooking disbenefits to ecosystems and less privileged people.

This new paradigm requires explicit recognition and response to the strengths but also the wider negative consequences of the weaknesses of all approaches, engineered and nature- and community-based alike, and to resolve them in wider thinking about hybridisation of techniques to reduce or mitigate unavoidable externalities.[101] This thinking has been applied in India to the Banas catchment in Rajasthan, at present in principle but with the intent of progressively influencing policy and practice focused on regeneration of the currently degraded functioning of the catchment system and its linked social and ecological dependents (Box 3.13). Whilst natural and traditional infrastructure generates a wide range of ecosystem services, it does so at limited capacity. Conversely, 'heavy engineering' approaches maximise selected services but with many externalities expressed across the catchment system. Restoration or innovation of groundwater recharge practices, particularly in the upper catchment, can represent a proven, ecosystem-based approach to resource regeneration with linked socio-ecological benefits for all dependents across the catchment. Explicit recognition of all potential benefits but also limitations and externalities of technologies can support the informed hybridisation of ecosystem-based with engineered methods, simultaneously increasing the security of all connected livelihoods and the vitality of the supporting river system. If, for example, water recharge practices were subsidised, supported by reinvestment of a premium on intensive water users benefitting from water transferred from the dam, self-beneficial

---

[100] World Commission on Dams. (2000). *Dams and Development: A New Framework for Better Decision-making*. Earthscan, London.
[101] Everard, M. (2013). *The Hydropolitics of Dams: Engineering or Ecosystems?* Zed Books, London.

outcomes for communities in the upper catchment would also support water security for those depending on abstractions lower in the catchment.[102] Policy reform focused on the functioning of the underpinning ecosystem, specifically emphasising water recharge, can thereby contribute to water security and yield socio-economic outcomes for all linked beneficiaries through a systemic understanding of how the water system functions. An approach founded on supportive ecosystem processes, rather than narrowly framed demands, can also serve to connect goals and budgets across multiple, currently fragmented policy areas.

Box 3.13 Challenges and Ecosystem-Focused Opportunities in Reanimating the Banas Catchment, Rajasthan

Sustainable solutions to water stewardship lie not so much in either engineering or nature-based approaches alone, but in their context-specific hybridisation supporting local, rural needs whilst replenishing ecosystems from which large-scale water resources are withdrawn. The Banas catchment in Rajasthan, India, exemplifies both the nature of the problem but also a quest for sustainable solutions, albeit being pursued in the face of strongly entrenched pro-engineering presumptions.[103]

The city of Jaipur had formerly subsisted on its own groundwater and monsoon recharge systems, but depleted and polluted these as the city grew. Municipal authorities reached out 32 km to the north-east to the Ramgarh Dam from 1952, appropriating resources intended for local uses and, after several phases of dam-raising, contributed substantially to the drying on the Ramgarh Lake since 2000. In 2006, City authorities then instigated a major transfer scheme appropriating waters from some 120 km to the south from the Bisalpur Dam on the Banas River. The Bisalpur Dam too had initially been built for local needs, and appropriation of its water occurred in the face of fierce local opposition in which some protestors were killed. The Bisalpur Dam now fails to fill reliably and the water is of

---

[102] Everard, M. (2019). A socio-ecological framework supporting catchment-scale water resource stewardship. *Environmental Science and Policy*, 91, pp. 50–59.

[103] Everard, M., Sharma, O.P., Vishwakarma, V.K., Khandal, D., Sahu, Y.K., Bhatnagar, R., Singh, J., Kumar, R., Nawab, A., Kumar, A., Kumar, V., Kashyap, A., Pandey, D.N. and Pinder, A. (2018). Assessing the feasibility of integrating ecosystem-based with engineered water resource governance and management for water security in semi-arid landscapes: A case study in the Banas Catchment, Rajasthan, India. *Science of the Total Environment*, 612, pp. 1249–1265.

declining quality, due substantially to the proliferation of tube wells in the upper catchment driving the abandonment of traditional communal monsoon recharge practices and also exposing people and croplands to geologically contaminated water.

Jaipur is faced with a dilemma. It may look even more remotely for inevitably contested water resources, following the sadly all too commonly followed "civil engineering paradigm". Pursuit of this flawed cycle of "taking more from further" is a demonstrably dysfunctional model, also omitting consideration of long-term ramifications that create a linked set of vulnerabilities for urban, rural, irrigation, wildlife and associated ecotourism beneficiaries alike in the Banas. Alternatively, authorities in Jaipur could pursue a strategy based on sustainable management of the water resources of the Banas system, as well as other sources feeding the city, by promoting catchment-scale thinking that subsidises or otherwise promotes resource regeneration to balance abstractions.

A major workshop in late 2017 brought together people from government, academia, NGOs and communities across the Banas catchment to address and seek progress with this vision.[104] A renewed policy focus on local-scale water recharge practices balancing water management and extraction technologies is consistent with emerging Rajasthani policies, particularly the visionary *Mukhya Mantri Jal Swavlamban Abhiyan* (MJSA: 'Chief Minister's Water Self-reliance Mission')[105] programme. At the present time, exploration of a "more from further" model seems to be the dominant paradigm in government thinking.

3.2   Reanimating Catchment Ecosystem Services in the Developed World

Rehydration of socio-ecological systems yields far broader benefits that water resource provision alone. Water resource security has served as an anchor service in the preceding exemplars, but a regenerative approach has also co-produced multiple linked beneficial outcomes. Equally, other desired outcomes—such soil retention and quality, nutrient cycling, habitat for biodiversity or water storage—can serve as the primary anchor

---

[104] Everard, M., Sharma, O.P., Singh, A.K., Vishwakarma, V.K. and Pinder, A.C. (2018). *Regeneration of the Banas-Bisalpur Socio-ecological Complex: Workshop Report, JKLU, Jaipur (4–6th December 2017)*. University of the West of England, Bristol.

[105] http://mjsa.water.rajasthan.gov.in/, accessed 8 January 2019.

service. Ecosystem-based landscape and waterscape management systemic solutions can co-produce a linked suite of ecosystem service benefits by increasing the capacity of landscapes to sustain human wellbeing. This section explores some of the wide range of other exemplars of ecosystem-based management achieving linked ecosystem service outcomes in the already-developed world.

3.2.1 Catchment Ecosystem Services for Water Quality

There is growing awareness of the potential for ecosystem-based landscape and waterscape management to protect or regenerate water storage and purification, soil retention and quality, nutrient cycling and biodiversity, and their multiple associated human benefits.

Three of a range of water resource-based global exemplars reviewed in more detail in my book *The Hydropolitics of Dams*[106] include the Upstream Thinking programme in south-west England (Fig. 3.7) (Box 3.14), New York City's water supply (Box 3.15) and SCaMP in north west England (see Box 3.16). All have shifted focus from downstream technical solutions towards upstream protection of ecosystem processes safeguarding raw water.

Box 3.14 'Upstream Thinking', South West England

The Upstream Thinking programme[107] in south west England, operated by South West Water (SWW, the regional water utility), reinvests a proportion of water service charges into improvements to agricultural practices in catchments serving surface water abstraction points.[108] By reducing inputs of particulate, soluble and microbial pollutants from farmed land, raw water quality is protected as part of SWW's aim to reduce chemical, financial and energy inputs to potable supply.

Business benefits accrue to SWW and its bill-paying customers. OFWAT (the economic regulator of the water services industry in England and Wales) accepted SWW calculations that Upstream Thinking represents a

---

[106] Everard, M. (2013). *The Hydropolitics of Dams: Engineering or Ecosystems?* Zed Books, London.

[107] www.upstreamthinking.org.

[108] Upstream Thinking. (n.d.). Upstream thinking. www.upstreamthinking.org, accessed 8 April 2015.

65:1 benefit-to-cost ratio relative to treatment of more contaminated water. However, although water quality is the 'anchor service', land management and other solutions applied also optimise multiple linked socio-ecological co-benefits for fisheries and river ecosystems, farm viability, wildlife and ecotourism.[109]

Box 3.15 New York City's Water Supply

New York City derives its water supply from Cat/Del (Catskills and Delaware) catchments. A financial contract was negotiated between urban water users and farming and other rural communities in the Cat/Del catchments to undertake measures to maintain high quality water. This has become one of the largest global 'payment for ecosystem service' (PES) schemes. This arrangement was formalised as a comprehensive Memorandum of Agreement in 1987. Under the terms of the Memorandum, the city committed funds of approximately $US350 million (£190 million) with additional investment in a watershed protection programme costing approximately $US1.3 billion (£700 million).

Though substantial, these figures represent only a small fraction of the financial costs and environmental impacts of alternative conventional engineered solutions to treat more contaminated raw water abstracted downstream. This partnership approach, linking rural and urban stakeholders, is key to maintaining New York City's pristine water quality and the viability of farming in source catchments for the foreseeable future.

Box 3.16 SCaMP, the Sustainable Catchment Management Programme, in North-West England

SCaMP, the Sustainable Catchment Management Programme,[110] was instigated by the British multi-utility company United Utilities (UU), the water service provider for the north west of England. UU has substantial upland landholdings to protect water quality and support significant habitats for a variety of wildlife.

SCaMP was developed in partnership with wildlife NGO the Royal Society for the Protection of Birds (RSPB). The first phase of SCaMP in 2005–2010

---

[109] South West Water. (2012). *Corporate Sustainability Report 2012*. https://www.southwestwater.co.uk/media/pdf/s/t/SWW-Corporate-Sustainability-Report-2012.pdf, accessed 8 April 2015.

[110] http://corporate.unitedutilities.com/cr-scamp.aspx.

entailed working with tenant farmers on UU-owned land to revise management practices and undertake additional capital works to restore upland
habitat simultaneously beneficial for water production, scarce ecosystems
and farm incomes, funded through reinvestment of water service charges.[111]
Subsequent phases of SCaMP have addressed water capture areas not
owned by UU, but where targeted subsidy and advice achieves water quality and quantity outcomes beneficial to the water company, its customers,
biodiversity and connected services.[112]

They all yield financial 'bottom line' benefits related to savings of water
treatment costs, but also multiple co-beneficial outcomes including fishery,
biodiversity and ecotourism protection, and stabilisation of farm incomes,
all of cumulatively substantial societal value.

3.2.2 Catchment Ecosystem Services for Flood Management

Within Europe, management of flood impacts is making a transition
from 'defence' of assets at risk towards a catchment-based approach to
retaining water upstream, potentially also recognising ecosystem service co-benefits from landscape-scale interventions.[113] This approach is
coalescing around the concept of 'natural flood management' (NFM),
defined as the alteration, restoration or use of landscape features as a
novel way of reducing flood risk.[114,115,116] The implementation and

---

[111] Everard, M. (2009). *The Business of Biodiversity*. WIT Press, Lyndhurst.

[112] United Utilities. (n.d.). SCaMP. http://corporate.unitedutilities.com/cr-scamp.aspx, accessed 3 April 2015.

[113] Everard, M., Bramley, M., Tatem, K., Appleby, T. and Watts, W. (2009). Flood management: From defence to sustainability. *Environmental Liability*, 2, pp. 35–49.

[114] Parliamentary Office of Science and Technology. (2011, December). *Natural Flood Management*. POSTNOTE 396. The Parliamentary Office of Science and Technology, HM Government, London.

[115] Morris, J., Hess, T.M., Gowing, D.G., Leeds-Harrison, P.B., Bannister, N., Vivash, R.M.N. and Wade, M. (2005). A framework for integrating flood defence and biodiversity in Washlands in England. *International Journal River Basin Management, IAHR and INBO*, 3(2), pp. 1–11.

[116] Wheater, H. and Evans, E. (2009). Land use, water management and future flood risk. *Land Use Policy*, 265, pp. 251–264.

Fig. 3.7 Installation of buffer zones to prevent cattle trampling and defecating in river margins and to attenuate pollutants in surface run-off are one of many solutions implemented under the 'Upstream Thinking' programme. (Image © Dr Mark Everard)

continued success of NFM schemes depends on effective collaboration between land-owners and communities, long-term funding measures or incentives, and better use of local knowledge (Fig. 3.8).

An impressive example of multi-partner NFM in action to address flood risk and generate a range of linked co-benefits is seen around the town of Stroud in the county of Gloucestershire in south west England

Fig. 3.8 Floodwater allowed to fill the floodplains it has sculpted in the land-scape is retained and slowed, reducing flood risks downstream. (Image © Dr Mark Everard)

(Box 3.17). However, a significant obstacle to NFM is that there is at present no enforceable policy or agreed framework to recognise and economically quantify the full spectrum of ecosystem service co-benefits that NFM schemes create. Nor are there appropriate legislation, policy instruments and associated budgets. This 'regulatory lag' between bold, multi-beneficial, cost-effective and sustainable aspirations and practical policy drivers highlights the lack of coherence between stated commitments and operational realities that often favour short-term commercial returns from land uses.[117]

---

[117] Everard, M., Dick, J., Kendall, H., Smith, R.I., Slee, R.W., Couldrick, L., Scott, M. and MacDonald, C. (2014). Improving coherence of ecosystem service provision between scales. *Ecosystem Services*, 9, pp. 66–74.

Box 3.17 The Stroud Rural SuDS Project

An impressive example of a multi-functional NFM is the Stroud Rural SuDS Project in sub-catchments upstream of the town of Stroud in the county of Gloucestershire, south-west England.[118] Using a diversity of measures to retain water and slow flows in the upper catchment, the Stroud Rural SuDS Project works in partnership with local community flood groups, farmers, partner organisations and major landowners including the National Trust (a UK heritage and nature conservation organisation holding lands in trust for public benefit[119]) to reduce flood risk in the lower River Frome valley and particularly in the town of Stroud. Effective landscape interventions include installing large natural woody material dams where safe and feasible, as well as water-holding depressions in fields designed so as do not adversely interfere with farming activities.

The 'anchor service' driving investment and collaboration is flood risk management. However, habitat-based measures also co-produce a wide range of additional ecosystem service benefits, including improvements in water quality, biodiversity and landscape aesthetics of the streams, brooks and wider catchment of the River Frome system. Partnership working and achievement of multiple linked ecosystem service benefits are integral to the programme.

3.2.3 Catchment Ecosystem Services for Resource Conservation and Wider Outcomes

Although the examples of Upstream Thinking, New York City's water supply and SCaMP all deliver multiple co-benefits, their driving purpose, or anchor service, is protection of raw water quality for subsequent extraction for public supply. However, other catchment-based approaches target broader suites of ecosystem service outcomes.

Two major US initiatives offer financial incentives for private farmland owners to optimise ecosystem service outcomes: the Conservation Reserve Program (Box 3.18) and measures to reduce nutrient loadings to Chesapeake Bay (Box 3.19). In both examples, the land remains in pri-

---

[118] Stroud District Council. (n.d.). *The Stroud Rural SuDS Project*. Stroud District Council. https://stroud.gov.uk/docs/environment/rsuds/index.asp, accessed 15 March 2016.
[119] National Trust. (n.d.). *The National Trust*. https://www.nationaltrust.org.uk/, accessed 15 March 2016.

vate ownership with rewards to owners routed through regulatory bodies on behalf of the general public benefiting from enhanced ecosystem functioning. These innovative policy and economic models reward targeted improvements to ecosystem functioning, delivering wider public benefits.

Box 3.18 A PES Approach to Optimising Public Benefits from Private US Farmland

The US Conservation Reserve Program (CRP) is a large-scale 'payment for ecosystem services' (PES) scheme, instituted in 1985 to avert soil erosion but evolving to address a broader 'bundle' of linked ecosystem services. The CRP operates as a land set-aside programme with government, via the US Department of Agriculture (USDA), paying landowners incentives through 10- to 15-year contracts to change the use of specific lots of land for ecosystem service benefits.[120]

Contracts are now let through 'reverse auctions' in which potential ecosystem service 'sellers' submit bids indicating the minimum payment they are willing to accept for practices delivering desirable ecosystem service. Bids are prioritised by the USDA according to potential environmental benefits, integrating these with proposed payments into an Environmental Benefit Index (EBI) and awarding contracts on the basis of the highest service benefit for least cost. Without CRP payments, 51% of CRP land would be returned to crop production with a corresponding decline in expenditure on outdoor recreation worth up to $300 million annually in rural areas.[121]

Box 3.19 An Economic Approach to Reducing Nutrient Inputs to Chesapeake Bay

Nutrient enrichment is a major threat to the economically- and environmentally-important Chesapeake Bay in the north-east USA. An innovative, market-based scheme operates through a programme of subsides for measures to reduce nutrient loadings to the Bay from upstream

---

[120] OECD. (2010). *Paying for Biodiversity: Enhancing the Cost-Effectiveness of Payments for Ecosystem Services*. OECD, Paris.

[121] Sullivan P., Hellerstein, D., Hansen, L., Johansson, R., Koenig, S., Lubowski, R., McBride, R., McGranahan, D., Roberts, M., Vogel, S. and Bucholtz, S. (2004). *The Conservation Reserve Program: Economic Implications for Rural America*. Agricultural Economic Report, Vol. 834. USDA Economic Research Service. http://www.ers.usda.gov/publications/aer834/aer834.pdf, accessed 2 May 2011.

farmland. It constitutes a 'payment for ecosystem services' (PES) approach, innovatively tackling water management problems where traditional regulation has failed.[122]

As point wastewater sources contribute only a fraction of total load of nutrients entering the Bay, Watershed Implementation Plans (WIPs) focus on programs to limit diffuse and weather-dependent loads from agriculture and low-density suburban development. Published total daily maximum load (TMDL) calculations for nitrogen, phosphorus and fine sediment for the Bay are fed into WIPs for six states and Washington DC as a basis for targeting diffuse pollution controls.

Farming interests bid for how much they are willing to accept in payment to implement pre-approved best management practice (BMP) 'Bay-friendly' nutrient reduction measures. Regulators then allocate subsidies under multi-year contracts where they will achieve the greatest net benefit per unit investment through a 'reverse auction' process.

3.3   Trees, Water and Livelihoods

The significant role of trees in water cycle and landscape rehabilitation is an element of some case studies already addressed. Forest ecosystems provide habitat for more than half the world's terrestrial species and are vital for carbon and other biogeochemical cycling systems. Approximately 240 million people, many of them lacking land tenure, live in the world's forests.[123] Yet all in global society benefit from the services of forests, as forested catchments are the source of more than three-quarters of the world's accessible freshwater, also supplying diverse other ecosystem services essential for human wellbeing at scales from the global to the local.[124]

---

[122] Ator, S.W. and Denver, J.M. (2015). *Understanding Nutrients in the Chesapeake Bay Watershed and Implications for Management and Restoration—The Eastern Shore (ver. 1.2, June 2015)*. US Geological Survey Circular 1406, US Geological Survey Circular, 72pp. https://doi.org/10.3133/cir1406.

[123] World Bank. (2003). *World Development Report 2003: Sustainable Development in a Dynamic World: Transforming Institutions, Growth, and Quality of Life*. World Bank, Washington, DC.

[124] Shvidenko, A., Barber, C.V., Persson, R., Gonzalez, P., Hassan, R., Lakyda, P., McCallum, I., Nilsson, S., Pulhin, J., van Rosenburg, B. and Scholes, R. (2005). Forest and woodland systems. In: *Millennium Ecosystem Assessment—Ecosystems and Human Well-being: Current State and Trends*, Chap. 21, pp. 585–621. http://www.millenniumassessment.org/documents/document.290.aspx.pdf, accessed 7 March 2016.

At continental scale, forests are efficient in the recycling of water, generating much of the rain that falls back into the forest ecosystem. They also act as 'continental water pumps', recycling water from moist coastal regions progressively further inland. Forest loss can therefore severely disrupt water balances at continental scale.[125] Forested uplands play particularly significant roles in intercepting moisture from oceanic air currents, generating the headwaters of river systems that irrigate whole subcontinents. Even small stands of forest recycle water efficiently in tight cycles, and can be seen creating lush and diverse ecosystems with cool microclimates in gully and valley ecosystems within arid landscapes.

Deforestation has generated many examples of hydrological disruption and adverse linked socio-ecological outcomes around the world. Conversely, forest management, conservation and, ideally, restoration can play significant roles in water and nutrient cycles, soil stabilisation, carbon sequestration, habitat for wildlife, landscape rehabilitation and provision of societal benefits often remote from where these services are produced. Forest protection and reforestation can therefore form a foundation for linked environmental and socio-economic rejuvenation, including halting former spirals of degradation.

3.3.1 Indigenous Forest Restoration on the Coromandel Coast

In the southern Indian state of Tamil Nadu, a familiar litany of landscape degradation is being driven by abandonment of traditional community management of tank (monsoon water interception and storage) systems[126,127] as well as substantial loss of natural forest cover, leading to severe land degradation and erosion. A long-term, dedicated approach to

[125] Aragão, L.E.O.C. (2012). Environmental science: The rainforest's water pump. *Nature*, 489, pp. 217–218. https://doi.org/10.1038/nature11485.

[126] Kajisa, K., Palanisami, K. and Sakurai, T. (2004). *Declines in the Collective Management of Tank Irrigation and Their Impacts on Income Distribution and Poverty in Tamil Nadu, India*. FASID Discussion Paper Series on International Development Strategies, No. 2004-08-005. Foundation for Advanced Studies on International Development, Tokyo.

[127] Kajisa, K., Palanisami, K. and Sakurai, T. (2007). Effects on poverty and equity of the decline in collective tank irrigation management in Tamil Nadu, India. *Agricultural Economics*, 36(3), pp. 347–362.

restoration of tropical dry evergreen forest (TDEF) for multiple beneficial outcomes has been occurring here for a number of decades. TDEF is
now a critically endangered habitat,[128] though once extensive along the
Coromandel Coast on the south-eastern seaboard of southern India.[129]
Though locally characteristic and of conservation importance,[130] TDEF
constitutes a biome rather than a distinctive forest type, shaped as much
by a long history of human pressures as natural forest succession.[131]
However, there is general consensus about the importance of restoring
TDEF for its broad associated biodiversity, medicinal, spiritual and other
uses and meanings, and a range of other beneficial ecosystem services.
Many remnant TDEF stands now form 'sacred groves' often associated
with Hindu temples, this sacred association having been influential in
their preservation. TDEF patches may also contain a diversity of medicinal plants used by local traditional healers[132] and that are also used by
local people for raft-making, haircare, religious purposes, fuelwood, edible fruits, pesticide, fodder and carpentry.[133]

Active restoration of TDEF has been occurring since 1973 at
Pitchandikulum Forest on the Auroville Plateau, Tamil Nadu. Starting
from a 65 acre (26.3 ha) site of severely degraded and eroded land of
negligible value, this restoration activity has demonstrated the feasibility
of eco-restoration with well-documented recovery of an increasing range
of native wildlife (including over 300 species of woody plants[134]).

---

[128] WWF. (n.d.). *Tropical and Subtropical Dry Broadleaf Forests—Southern Asia: Southern India*. Worldwide Fund for Nature. http://www.worldwildlife.org/ecoregions/im0204, accessed 3 March 2016.

[129] Blanchflower, P. (2005). Restoration of the tropical dry evergreen forest of peninsular India. *Biodiversity*, 6(3), pp. 17–24.

[130] Selwyn, M.A. and Udayakumar, M. (2008). Tropical dry evergreen forests of peninsular India: Ecology and conservation significance. *Tropical Conservation Science*, 1(2), pp. 89–110.

[131] Everard, M. (2018). The characteristics, representativeness, function and conservation impor-tance of tropical dry evergreen forest (TDEF) on India's Coromandel Coast. *Journal of Threatened Taxa*, 10(6), pp. 11760–11769.

[132] Parthasarathy, N., Selwyn, M.A. and Udayakumar, M. (2008). Tropical dry evergreen forests of peninsular India: Ecology and conservation significance. *Tropical Conservation Science*, 1(2), pp. 89–110.

[133] Kinhal, V. and Parthasarathy, N. (2008). Secondary succession and resource use in tropical fal-lows: A case study from the Coromandel Coast of South India. *Land Degradation and Development*, 19(6), pp. 649–662.

[134] Pitchandikulam Virtual Herbarium. http://www.pitchandikulam-herbarium.org/, accessed 28 May 2016.

Restoration of TDEF has seen an associated regeneration of herbal tradi-tions, forest-based livelihoods, and traditional and religious benefits such as enhancement of sacred groves.[135]

Nadukuppam Forest in the Kaliveli catchment of Tamil Nadu is the subject of ongoing TDEF restoration on formerly degraded and severely eroded farmland, initially on 30 acres (0.12 km^2) but with more in plan to restore and put into trust as part of a longer-term programme of eco-restoration. Connected with Naddukuppam Forest restoration are the adjacent 'Nadukuppam Field' and Nadukuppam School. Nadukuppam Field is a women's collective set of businesses including herbal medicinal products and crafts linked to forest products, local employment of women in germinating and nurturing indigenous forest plants, and production of the alga *Spirulina* as a dietary supplement. Nadukuppam School has an environmental programme permeating its cross-curricular learning, and is used as a model for many schools across Tamil Nadu state. Nadukuppam Forest eco-restoration therefore links with economic and educational development and women's empowerment (Fig. 3.9).

TDEF stores significant reserves of carbon above and below ground and in the underlying soil. One of the principal funding mechanisms supporting expansion of reforestation at Nadukuppam is an innovative developed-developing world partnership led by the NGO The Converging World (TCW), twinning aspirations for low-carbon development in south west England and Tamil Nadu.[136] Under this model, TCW invests funds from south west England in wind turbines in Tamil Nadu, provid-ing the joint benefits of 'offsetting' some emissions from the developed region of England simultaneously with promoting a lower-carbon path-way of development in Tamil Nadu. However, there is an additional 'multiplier effect' achieved through channelling part of the surplus income from renewable energy sales into the restoration of tracts of TDEF. Over the respective lifetime of the wind turbines (nominally 20

---

[135] Pitchandikulam Forest. http://www.pitchandikulamforest.org/PF/.
[136] Everard, M., Longhurst, J.W.S., Pontin, J., Stephenson, W. and Brooks, J. (2017). Developed-developing world partnerships for sustainable development (1): An ecosystem services perspective. *Ecosystem Services*, 24, pp. 241–252.

Fig. 3.9 Ladies from Nadakuppam village water saplings forming part of the restored Naddukuppam Forest. (Image © Dr Mark Everard)

years) and progress to maturity of the restored TDEF (assumed as 100 years), sequestration by TDEF substantially augments carbon dioxide emissions averted by low-carbon generation by turbines, more or less doubling the overall climate benefits and representing significant cost-efficiency accelerating multinational progress with low-carbon development.[137] Climate regulation is an anchor service[138] supporting the initial business case at Naddukuppam, but forest restoration also optimises a spectrum of ecosystem service outcomes from which tangible local benefits arise and for which future markets may eventually be developed. These diverse services are significant for local people, who are key players in maintaining the 'cultural landscapes' that support their needs.[139]

[137] Everard, M., Longhurst, J.W.S., Pontin, J., Stephenson, W., Brooks, J. and Byrne, M. (2018). Developed-developing world partnerships for sustainable development (3): Reducing carbon sequestration uncertainties in south Indian tropical dry evergreen forest. *Ecosystem Services*, 33, pp. 173–181.

[138] Everard, M. (2014). Nature's marketplace. *The Environmentalist*, March 2014, pp. 21–23.

[139] Schaich, H., Bieling, C. and Plieninger, T. (2010). Linking ecosystem services with cultural landscape research. *GAIA*, 19(4), pp. 269–277.

3.3.2 Emulating the Water Capture Role of Trees and Forests

The roles that trees and forests play in capturing water from the atmosphere are well known. Nature has many other innovations for capturing water from the atmosphere including, for example, the aerial roots on orchids that rapidly absorb water and nutrients from the air. Pine trees are amongst many species efficient at the capture of water from moist air, and aerial moisture can have a bigger effect on forest productivity than moisture in the soil.[140] Animals too can perform this apparent conjuring trick of drawing water from thin air. For example, the Namibian fog beetle (*Stenocara gracilipes*) is also known as the fogstand beetle, adapted to southern Africa's Namib Desert, one of the most arid areas of the world receiving only 1.4 centimetres of rain per year. The ridges of the beetle's shell constitute the coolest surface at night and condense water into tight droplets such that, in the morning, the beetle tips up its shell to allow the water to run down to its mouth.

The role of forests in moisture interception and retention is perhaps unfortunately best evidenced by cases where forest clearance has changed weather and local climate patterns. However, there are many more local uses of this ecosystem service. As one example, nomadic tribes in Saudi Arabia obtain clean water by placing cisterns beneath trees known to trap water from moist air during periods of intense fog.

This moisture-trapping phenomenon has been emulated technologically in the form of fog and mist capture nets in Saudi Arabia, Chile, Peru and Ethiopia (see Box 3.20). These methods are of course not nature itself, but learn from natural processes (an idea we will return to later in this chapter when considering 'biomimicry'). Such nature-emulating approaches to water harvesting may prove important in serving the needs of people in some arid coastal areas as well as upland zones where natural water-capturing habitats have become degraded. Moisture-trapping may also potentially act as a temporary means to restore hydrology in formerly wooded areas, which may then be reforested to subsequently resume their natural water harvesting function amongst other benefits to society.

---

[140] Wei, L., Zhou, H., Link, T.E., Kavanagh, K.L., Hubbart, J.A., Du, E., Hudak, A.T. and Marshall, J.D. (2018). Forest productivity varies with soil moisture more than temperature in a small montane watershed. *Agricultural and Forest Meteorology*, 259, pp. 211–221.

Box 3.20 Examples of Mist and Fog Water Capture Schemes

In Asir, south western Saudi Arabia, large nets have been deployed since 2006 to intercept fog, water droplets coalescing on the netting as an alternative source of fresh water in this dry region.[141]

Fog-trapping along the ridge of the 2,600-foot El Tofo mountain in Chile operated between 1987 and 2002.[142] Arrays of plastic nets harvested moisture from clouds blowing in from the Pacific Ocean, collecting droplets and piping them down seven kilometres to serve the fishing village of Chungungo. Though cheap and low-maintenance, political decisions discontinued the programme in 2002.

The Atacama Desert is the driest place on Earth, rainfall having never been recorded. Much of the Atacama Desert is in Chile, where polypropylene 'fog catcher' nets harvest moisture that local people collect for drinking and agriculture.[143] There is a hope that this may offer the potential for transforming the desert into large-scale farms and gardens.

Fog catchers dot the deserts around Lima, Peru, the world's second-largest desert city after Cairo with a population of 9 million. They provide a significant water source augmenting rainfall of less than 4 cm annually, erratic flows of melt water from dwindling Andean glaciers and transfers from far-off Andean lakes.[144] In winter, dense fog sweeps in from the Pacific Ocean and, for an up-front cost of a few thousand dollars and some volunteer labour, village can set up fog-collecting nets that gather hundreds of gallons of water a day.[145]

In Ethiopia, WarkaWater towers are synthetic structures, inspired by the Warka tree that harvests atmospheric water vapour.[146] WarkaWater towers comprise a semi-rigid exoskeleton built from locally available reeds and bamboo with an internal polymer mesh to trap condensation, offering a durable and environmentally, financially and socially sustainable solution.

---

[141] Al-hassan, G.A. (2009). Fog water collection evaluation in Asir Region–Saudi Arabia. *Water Resource Management*, 23, pp. 2805–2813.

[142] FogQuest. (2009). *Chile—El Tofo/Chungungo 1987—2002*. FogQuest. http://www.fogquest. org/project-information/projects/chile-el-tofo-chungungo/, accessed 6 June 2016.

[143] McCann, J. (2013). Bringing the driest place in the world to life: 'Fog catchers' attempt to harvest moisture with huge nets in Chilean desert. *Mail Online*, 30 March 2013. http://www.dailymail.co. uk/news/article-2301226/Fog-catchers-attempt-harvest-moisture-huge-nets-Chilean-desert.html, accessed 6 June 2016.

[144] Collyns, D. (2012). Peru's fog catchers net water supplies. *The Guardian*, 19 September 2012. http://www.guardian.co.uk/global-development/2012/sep/19/peru-fog-catchers-water-supplies, accessed 6 June 2016.

[145] Fields, H. (2009). Fog catchers bring water to parched villages. *National Geographic*, 9 July 2009. http://news.nationalgeographic.com/news/2009/07/090709-fog-catchers-peru-water-missions.html, accessed 6 June 2016.

[146] Flaherty, J. (2014). A giant basket that uses condensation to gather drinking water. *Wired*, 28 March 2014. http://www.wired.com/2014/03/warka-water-africa/, accessed 6 June 2016.

3.3.3   Other Forest- and Tree-Related Initiatives Underpinning Socio-Ecological Regeneration

Forests, individual trees and small stands produce a multiplicity of eco-system services. It is then perhaps unsurprising that there are many other examples, from countries across a diversity of biogeographical zones and states of development, where forest protection or enhancement schemes have revived linked environmental and societal wellbeing.

A subset of developing world examples include the Miaro forest corridor project in Madagascar (Box 3.21) and the Pagos por Servicios Ambientales (PSA) scheme in Costa Rica (Box 3.22)

Box 3.21 The Miaro Forest Corridor Project, Madagascar

The *Miaro* project, undertaken by the NGO WWF (the World Wide Fund For Nature) in the humid Fandriana-Vondrozo forest corridor, Madagascar, is based on 'payment for ecosystem services' (PES) principles. Payments from downstream beneficiaries of water supplies emanating from high biodiversity forest sites are recirculated to upstream communities to reward them for conservation and sustainable management of the watershed.[147]

The water supply emanating from upland forests serves as an economically valuable anchor service for the PES. However, forest conservation also yields many linked, largely non-market ecosystem services, including the nature conservation values of upland humid forest and livelihood improvements for the rural poor.

Box 3.22 Pagos por Servicios Ambientales (PSA), Costa Rica

In Costa Rica, Central America, the Pagos por Servicios Ambientales (PSA: Payments for Environmental Services) scheme has operated since 1996 to provide economic incentives for forest conservation. PSA directs payments to ecosystem service outcomes generated by forest and agro-forestry eco-

---

[147] WWF. (2009). *Watershed-based Payments for Ecosystem Services in the Humid Forest of Madagascar.* Worldwide Fund for Nature. http://wwf.panda.org/who_we_are/wwf_offices/madagascar/?uProje ctID=MG0921#, accessed 7 March 2016.

systems, replacing a former ineffective system of tax deductions to support poorly-targeted forest conservation.[148,149]
Landowners entering PSA scheme are paid for four land use activities (protecting natural forest, establishing timber plantations, regenerating natural forest and establishing agro-forestry systems) producing a bundle of ecosystem services.[150] The scheme is funded by reallocation of 3.5% of the revenues from fossil fuel sales taxes, topped up by contributions from the World Bank and other international donors.[151] Individual beneficiaries (hydroelectric plants, breweries, irrigated farms and other organisations benefiting from ecosystem services) can also pay into the scheme, negotiating contracts with service providers.

Selected examples of forest protection or enhancement for combined socio-ecological benefit, amongst many in already-developed nations, include New Zealand's Nga Whenua Rahui conservation reserve programme (Box 3.23) and the forest restoration aspiration within the UK's 25-year environment plan[152] (Box 3.24).

Box 3.23 Nga Whenua Rahui Conservation Reserve Programme, New Zealand

New Zealand has implemented novel forest conservation schemes. Some work with indigenous Maori landowners in North Island who are interested in receiving payments for maintaining their land to preserve livelihoods and culture.[153] A Maori conservation reserve program known as Nga

---

[148] Food and Agricultural Organization of the United Nations. (2007). *The State of Food and Agriculture 2007: Paying Farmers for Environmental Services.* http://www.fao.org/publications/sofa/2007/en/ accessed 1 May 2011.

[149] Pfaff, A., Kerr, S., Lipper, L., Cavatassi, R., Davis, B., Hendy, J. and Sanchez, A. (2007). Will buying tropical forest carbon benefit the poor? Evidence from Costa Rica. *Land Use Policy*, 24(3), pp. 600–610.

[150] Wünscher, T., Engel, S. and Wunder, S. (2006). Payments for environmental services in Costa Rica: Increasing efficiency through spatial differentiation. *Quarterly Journal of International Agriculture*, 45(4), pp. 319–337.

[151] OECD. (2010). *Paying for Biodiversity: Enhancing the Cost-Effectiveness of Payments for Ecosystem Services.* OECD, Paris.

[152] HM Government. (2017). *A Green Future: Our 25 Year Plan to Improve the Environment.* HM Government, London. https://assets.publishing.service.gov.uk/government/uploads/system/uploads/attachment_data/file/693158/25-year-environment-plan.pdf, accessed 10 January 2019.

[153] Funk, Jason. (2006). Maori farmers look to environmental markets in New Zealand. *Ecosystem Marketplace*, 24 January 2006. http://www.ecosystemmarketplace.com/pages/dynamic/article.page.php?page_id=4097&section=home&eod=1.

Whenua Rahui provides economic support enabling landowners to allow land to remain in, or revert to, native bush.[154]
Nga Whenua Rahui is funded by government to reflect wider ecosystem service benefits to offset the impacts of New Zealand's rapidly urbanising economy. Some other government incentives also support carbon sequestration and management of erodible land.

Box 3.24 Forest Restoration Within the UK Government's 25-Year Plan

In 2015, the UK's Natural Capital Committee (NCC) recognised that 'natural capital deficits' are costly to societal wellbeing and the economy.[155] The NCC consequently recommended that government implement a 25-year environment plan, outlining an economic case for investment in habitat creation and restoration. This included planting 250,000 additional hectares of woodland, optimally located in the landscape for ecosystem service delivery. Economic returns were calculated to be at least as great as those from investment in traditional engineered infrastructure.

UK Government endorsed the NCC's proposed 25-year environment plan,[156] leading to publication in 2017 of the government's strategy document *A Green Future: Our 25 Year Plan to Improve the Environment*.[157] In *A Green Future*, there is a published intent to support development of a northern forest and 'zero-deforestation supply chains', recognising that England's woods and forests are a national asset delivering an estimated £2.3 billion annually in ecosystem services. At more local scale, the multiple benefits provided by street trees are also explicitly recognised and accorded more protection in the 25-year environment plan.

Globally, forests and wetlands are significant sinks of carbon. Many of the densest, as yet economically unexploited deposits are laid down in the forests

---

[154] Nga Whenua Rahui. (n.d.). *Nga Whenua Rahui.* http://www.doc.govt.nz/ngawhenuarahui, accessed 28 May 2016.

[155] Natural Capital Committee. (2015). *Protecting and Improving Natural Capital for Prosperity and Wellbeing: Third 'State of Natural Capital' Report.* Natural Capital Committee, HM Government, London. https://www.naturalcapitalcommittee.org/, accessed 21 November 2016.

[156] HM Government. (2015). *The Government's Response to the Natural Capital Committee's Third State of Natural Capital Report,* September 2015. https://www.gov.uk/government/uploads/system/uploads/attachment_data/file/462472/ncc-natural-capital-gov-response-2015.pdf, accessed 21 November 2016.

[157] HM Government. (2017). *A Green Future: Our 25 Year Plan to Improve the Environment.* HM Government, London. https://assets.publishing.service.gov.uk/government/uploads/system/uploads/attachment_data/file/693158/25-year-environment-plan.pdf, accessed 10 January 2019.

of developing countries. Deforestation and forest degradation account for nearly 20% of global greenhouse gas emissions, exceeding emissions from the global transportation sector and second only to the energy sector. Climate stabilisation is practically impossible without reducing emissions from forests. Forest loss and degradation is driven largely by agricultural expansion, conversion to pastureland, infrastructure development, destructive logging, fires and other causes. Much of the wealth of developed nations is founded on a historic climate-destructive pathway of development, creating moral as well as legal barriers to demanding that developing nations curb their aspirations for economic emancipation. Novel economic schemes, significantly including REDD+ (Box 3.25), can provide economic incentives for developing nations to retain carbon deposits in situ.

Box 3.25 The UN REDD+ Programme

REDD+ (the United Nations Collaborative Programme on Reducing Emissions from Deforestation and Forest Degradation in Developing Countries) creates financial value for carbon stored in forests, such that developing countries can invest in protection of forested lands as part of a low-carbon path to sustainable development.[158]

REDD+ addresses deforestation, forest degradation, conservation, sustainable management and enhancement of carbon stocks. It does so through a variety of mechanisms that open up market devices for payments for emission offsets by industrialised nations. These reward developing countries for protecting ecosystems of value for carbon storage and linked ecosystem service benefits.

Given the global importance of forest ecosystem services, the concept of sustainable forest management has been widely embraced at national and international policy levels, albeit that practical implementation is slow with little substantive progress yet made to address forest degradation globally.[159] However, the importance of forests for linked human

---

[158] UN-REDD Programme. (n.d.). *About REDD+*. UN-REDD Programme. http://www.un-redd.org/AboutREDD/tabid/102614/Default.aspx, accessed 28 May 2016.

[159] Shvidenko, A., Barber, C.V., Persson, R., Gonzalez, P., Hassan, R., Lakyda, P., McCallum, I., Nilsson, S., Pulhin, J., van Rosenburg, B. and Scholes, R. (2005). Forest and woodland systems. In: *Millennium Ecosystem Assessment—Ecosystems and Human Well-being: Current State and Trends*. Chap. 21, pp. 585–621. http://www.millenniumassessment.org/documents/document.290.aspx.pdf, accessed 7 March 2016.

wellbeing and biodiversity conservation has been recognised and promoted by a range of international agreements which aim to halt, and
ultimately reverse, forest loss and degradation (see Box 3.26).

Box 3.26 International Forest Protection and Regeneration Initiatives

The New York Declaration on Forests,[160] signed at the 2014 UN Climate Summit, is a non-binding global pledge endorsed by dozens of governments, thirty of the world's biggest companies and more than fifty influential civil society and indigenous organisations. The Declaration aims to restore 350 million hectares of deforested and degraded landscapes by 2030, and to cut natural forest loss in half by 2020 and end it by 2030. Halting forest loss would cut between 4.5 and 8.8 billion tons of carbon remobilised annually, approximating emissions by the United States. Commitments to landscape restoration under the New York Declaration on Forests include: Ethiopia (15 million hectares, or about five times the area of Belgium); Uganda (2.5 million hectares); the Democratic Republic of the Congo (8 million hectares); Colombia (1 million hectares); Guatemala (1.2 million hectares); and Chile (100,000 hectares). Many nations are expected to follow with their own commitments, with restoration of degraded land also likely to qualify for carbon credits.

The Bonn Challenge was established at a ministerial roundtable in September 2011 at the invitation of the German Government and IUCN. This Challenge calls for restoration of 150 million hectares (a little more than one-sixth the area of Australia) of deforested and degraded lands by 2020, facilitating implementation of existing international commitments requiring such restoration.[161] 150 million hectares of forest could sequester an additional 1 $GtCO_2e$ per year, a significant contribution to cutting global climate-active gas emissions, restoring ecological integrity and improving human wellbeing. Many governments, private companies and community groups have signalled their intent to align with the Bonn Challenge.

The Brazilian Coalition on Climate, Forests and Agriculture was initiated in 2014 as a mechanism spanning disciplines and sectoral interests cooperating in the conservation and sustainable use of forests, sustainable agriculture, and mitigation and adaptation to climate change in Brazil and

---

[160] United Nations. (2014). *FORESTS New York Declaration on Forests Action Statements and Action Plans.* http://www.un.org/climatechange/summit/wp-content/uploads/sites/2/2014/09/FORESTS-New-York-Declaration-on-Forests.pdf, accessed 8 April 2015.

[161] IUCN. (2016). *The Bonn Challenge.* IUCN. http://www.bonnchallenge.org/, accessed 18 March 2016.

worldwide.[162] Following climate change agreements achieved at the 2015 UNFCCC Conference of Parties in Paris, the Coalition is deepening the dialogue between its various participant sectors to promote sustainable development and a low-carbon economy to 2030.

These examples of forest protection or regeneration for ecosystem service enhancement, spread across a representative diversity of geographical zones, are a subset of schemes around the world. They highlight how priority policy areas—water, biodiversity, carbon, tribal lifestyles and many more—comprise different interlinked benefits flowing from regenerated forest lands. The policy priority driving change should be treated as a central anchor service, around which optimisation of other benefits may be planned taking a multi-service/outcome systemic solutions approach. Substantial knowledge transfer and targeting of global aid is available to support stated national commitments to restore forests, for example under PROFOR (Box 3.27).

Box 3.27 The Program on Forests (PROFOR)

The Program on Forests (PROFOR) multi-donor partnership, housed at the World Bank, generates innovative, cutting-edge knowledge and tools to advance sustainable management of forest resources for poverty reduction, economic growth, climate mitigation and adaptation, and conservation benefits.[163] PROFOR responds to the mounting pressure on forests worldwide, largely due to changing land uses and variations in weather patterns induced by global warming. This is occurring in the face of solid scientific consensus that deforestation, forest degradation, extended droughts, forest fires and rising temperatures will push key biomes, such as the Amazon, to reach tipping points in the near future resulting in the catastrophic collapse of vital services such as water and climate regulation. PROFOR is seeking to catalyse reforestation, responding to growing demands from governments for actionable information about how to maintain and build forest resilience, how and where to restore degraded forests, and how communities and economies can benefit from better forest management.

---

[162] Brazilian Coalition on Climate, Forests and Agriculture. (n.d.). *Brazilian Coalition on Climate, Forests and Agriculture.* http://www.coalizaobr.com.br/en/index.php?dm_i=2GI3,OLYJ,48BVGE,1MJYB,1, accessed 18 March 2016.

[163] PROFOR. (2017). *Know-How for Resilient Forest Landscapes: 2017 Annual Report.* Program on Forests (PROFOR).

PROFOR's work informs implementation of the World Bank Group Forest Action Plan (FAP) to help client countries establish resilient and sustainable forest landscapes that contribute to reducing poverty and increasing shared prosperity in a sustainable manner. PROFOR has played a key role in growing the portfolio of World Bank and other financial commitments as well as in the transfer of knowledge shaping a 'forest-smart' approach.

As one example, PROFOR is helping the government in Jamaica to assess the economic value of coastal protection services provided by mangroves, including their roles in buffering hurricanes and floods. In Myanmar, a PROFOR forest sector assessment initiated a process for identifying opportunities for forest and mangrove restoration to help reduce ongoing conflict over natural resources and to improve social inclusion, further informing the Government of Myanmar's program for large-scale restoration and reforestation. The World Bank has more than 20 mangrove projects underway globally, supported by PROFOR and initiated through various entry points (anchor services) such as disaster risk management, climate change, fisheries, agriculture and urban development.

A further important aspect of forest restoration is the need to not only decarbonise the economy—in other words to radically reduce societal carbon emissions—but also to decarbonise the atmosphere itself. Even were societal carbon emissions miraculously halted overnight, the amount of carbon dioxide already in the global atmosphere is contributing to a potentially dangerous degree of warming. Trees have a uniquely potent capability to help society make progress with negative carbon emissions, sequestering large volumes of carbon from the atmosphere. A study published in 2019 suggested that planting 0.9 billion hectares of trees (approximately the area of the USA) across the world could reduce global CO_2 concentrations by 25%, helping reduce risks of 'runaway' climate change.[164] Counterintuitively, increasing forest cover may become more feasible as a greater proportion of the growing global human population becomes urban, reliant on remaining outlying ecosystems for food but also a range of ecosystem services to which forest cover might contribute. The proposed solution of restoring global forests at unprecedented scale as *"the best climate change solution available"* may or may not be feasible, but would certainly be ineffective without radical cuts in anthropogenic emissions.

[164] Bastin, J.-F., Finegold, Y., Garcia, C., Mollicone, D., Rezende, M., Routh, D., Zohner, C.M. and Thomas W. Crowther, T.W. (2019). The global tree restoration potential. *Science*, 365(6448), pp. 76–79.

3.3.4   Restoration of Ecosystem Services Through Tree Removal

Forest restoration can make significant positive contributions to regenerative socio-ecologically beneficial landscapes. However, invasive vegetation, particularly invasive alien plants (IAPs) with greater water demands than locally adapted native species, may not only threaten biodiversity but also water security and linked ecosystem services including food production, soil erosion and fire risk, with poorer people suffering asymmetrically high impacts.[165] In Australia, it has been found that evaporative loss from one hectare of willows equates to water use by 17 households.[166] Consequently, IAP removal can improve water yields, reducing impacts on a range of beneficial ecosystem services.[167] Removal of invasive trees and protection of native forests may therefore be significant for water supply, flow regulation and other ecosystem services and associated livelihoods in heavily invaded tropical forests worldwide.[168] Notwithstanding the general global tendency of increasing invasion, there are significant regional successes in tackling IAPs for societal benefits.

South Africa is water-scarce, and particularly vulnerable to climate change and unsympathetic land uses. Widespread invasive alien plants, particularly trees, threaten the quality and quantity of water running off the landscape due to their greater evaporative loss compared to dryland-adapted species.[169] Species physiology, especially species-specific evapotranspiration rates in different environments,

---

[165] Asbjornsen, H., Tomer, M.D., Gomez-Cardenas, M., Brudvig, L.A., Greenan, C.M. and Schilling, K. (2007). Tree and stand transpiration in a Midwestern bur oak savanna after elm encroachment and restoration thinning. *Forest Ecology and Management*, 247, pp. 209–219.

[166] Doody, T.M. and Benyon, R.G. (2011). Quantifying water savings from willow removal in Australian streams. *Journal of Environmental Management*, 92, pp. 926–935.

[167] van Wilgen, B.W., Reyers, B., Le Maitre, D.C., Richardson, D.M. and Schonegevel, L. (2008). A biome-scale assessment of the impact of invasive alien plants on ecosystem services in South Africa. *Journal of Environmental Management*, 89(4), pp. 336–349.

[168] Cavaleri, M.A., Ostertag, R., Cordell, S. and Sack, L. (2014). Native trees show conservative water use relative to invasive trees: Results from a removal experiment in a Hawaiian wet forest. *Conservation Physiology*, 2(1), p. cou016. https://doi.org/10.1093/conphys/cou016.

[169] Dye, P. and Jarmain, C. (2004). Water use by black wattle (*Acacia mearnsii*): Implications for the link between removal of invading trees and catchment streamflow response. *South African Journal of Science*, 100, pp. 40–44.

plays a major role in the likely impacts of IAP species on water resources.[170] Rooting depth plays a key role in reducing water recharge.[171] Water security is one of the primary negative impacts associated with many IAPs in South Africa, particularly trees such as Australian *Eucalyptus* and *Acacia* species that have far greater evaporative loss compared to native species.[172] IAPs are conservatively estimated to use 2.9% of mean annual runoff in South Africa,[173] with reductions of more than 25% in many catchments and a likelihood of increasing reductions if IAPs are allowed to spread unchecked.[174] A particular effective example of invasive tree removal for water security and linked ecological and societal benefits is the Working for Water (WfW) programme in South Africa, operating since 1995 as a resource protection and employment programme administered through the Department of Public Works with the support of multiple government departments (Fig. 3.10). WfW is the largest and most effective such programme globally (Box 3.28).

Box 3.28 South Africa's Working for Water (WfW) Programme

Instigation of the 'Working for Water' programme in South Africa in 1995 was informed by the costs of reducing water loss by invasive species management compared with building new engineered infrastructure. WfW operates by providing employment for the least advantaged in society in the control of problematic invading alien plants. It is funded by multiple

---

[170] Le Maitre, D.C., Gush, M.B. and Dzikiti, S. (2015). Impacts of invading alien plant species on water flows at stand and catchment scales. *AoB Plants*, 7, p. plv043. https://doi.org/10.1093/aobpla/plv043.

[171] Seyfried, M.S. and Wilcox, B.P. (2006). Soil water storage and rooting depth: Key factors controlling recharge on rangelands. *Hydrological Processes*, 20, pp. 3261–3275.

[172] Dye, P. and Jarmain, C. (2004). Water use by black wattle (Acacia mearnsii): Implications for the link between removal of invading trees and catchment streamflow response. *South African Journal of Science*, 100, pp. 40–44.

[173] Le Maitre, D.C., Forsyth, G.G., Dzikiti, S. and Gush, M.B. (2016). Estimates of the impacts of invasive alien plants on water flows in South Africa. *Water SA*, 42(4), pp. 659–672.

[174] van Wilgen, B.W., Reyers, B., Le Maitre, D.C., Richardson, D.M. and Schonegevel, L. (2008). A biome-scale assessment of the impact of invasive alien plants on ecosystem services in South Africa. *Journal of Environmental Management*, 89(4), pp. 336–349.

government departments with interconnected aims of enhancing water security, ecological integrity, the productive potential of land, the quality of life of marginalised sectors of society through job creation and poverty alleviation, and economic benefits from wood, land, water and trained people.[175,176]

WfW is one of the biggest nature, water conservation, pro-poor employment and training programmes in the world. It has cleared more than one million hectares of IAPs through labour-intensive, mechanical, chemical, biological and integrated control, providing jobs and training to approximately 20,000 people annually from the most marginalised sectors of society.[177] Delivery of WfW is also linked with other socio-development initiatives, such as health and education. WfW's outcomes are social and environmental, generating multiple benefits associated with improved water yields[178,179] and linked ecosystem services.[180]

[175] Woodworth, P. (2006). Working for water in South Africa: Saving the world on a single budget? *World Policy Journal*, Summer 2006, pp. 31–43.

[176] Turpie, J.K., Marais, C. and Blignaut, J.N. (2008). The working for water programme: Evolution of a payments for ecosystem services mechanism that addresses both poverty and ecosystem service delivery in South Africa. *Ecological Economics*, 65(4), pp. 788–798.

[177] Department of Environmental Affairs. (n.d.). *Working for Water (WfW) Programme*. Department of Environmental Affairs, Government of South Africa. https://www.environment.gov.za/projectsprogrammes/wfw, accessed 30 May 2016.

[178] Le Maitre, D.C., Richardson, D.M. and Chapman, R.A. (2004). Alien plant invasions in South Africa: Driving forces and the human dimension. *South African Journal of Science*, 100, pp. 103–112.

[179] Marais, C., van Wilgen, B.W. and Stevens, D. (2004). The clearing of invasive alien plants in South Africa: A preliminary assessment of costs and progress. *South African Journal of Science*, 100, pp. 97–103.

[180] van Wilgen, B.W., Reyers, B., Le Maitre, D.C., Richardson, D.M. and Schonegevel, L. (2008). A biome-scale assessment of the impact of invasive alien plants on ecosystem services in South Africa. *Journal of Environmental Management*, 89(4), pp. 336–349.

Fig. 3.10 Eucalyptus trees tap water from deep aquifers and are 'thirsty' compared to native dryland-adapted trees in South Africa and India. (Image © Dr
Mark Everard)

3.4   Reanimating Entire Landscapes

Restoration of natural landscape functions is a linking theme of prior case
studies addressing water, catchment, forest and linked socio-economic
regeneration. In this section, we explore global exemplars of reanimated

landscapes over broader scales, integrally interlinking with rehabilitation of water and nutrient cycling, biodiversity, and the retention and formation of soils as a foundation for regenerating linked ecosystem and human wellbeing. The term 'reanimating' is used here as these examples embody far more than a traditional but narrow 'restoration' ethos, working instead with the foundational natural processes underpinning ongoing environmental and human vitality.

Major rehabilitation projects in China's Loess Plateau (Box 3.29) and the Ethiopian Highlands (Box 3.30) demonstrate that very large landscape-scale regeneration of the socio-ecological system is possible with vision, integrated policy, funding and involvement of local people, with major contributions to alleviating poverty by reversing former cycles of degradation.

Box 3.29 Restoring Socio-ecological Viability in China's Loess Plateau

Severe erosion resulting from intensive farming on sloping lands had formerly threatened the ecological integrity and socio-economic viability of the Loess Plateau in China's north-west, home to 50 million people. Centuries of over-use and over-grazing had created one of the highest erosion rates in the world, and a consequent negative spiral of socio-ecological decline and poverty.

To halt and reverse loss of the powdery loess soil that gives the Loess Plateau its name, the World Bank co-sponsored two major targeted restoration projects: the *Loess Plateau Watershed Rehabilitation Project* and the *Second Loess Plateau Watershed Rehabilitation Project*.[181] This ambitious, landscape-scale restoration sought to regenerate functional ecosystems, supporting sustainable agricultural production and viable livelihoods. The introduction of zoned grazing, terraced agriculture on slopes to protect soil, water and nutrients, controlled fuel wood gathering, and other forms of locally adapted, sustainable farming practices have doubled the coverage of perennial vegetation.

These measures have also doubled farmer incomes, enabling diversified employment and the production of a wider range of high-value products and greater productivity through creation of conditions for sustainable soil and water conservation. Food supplies have also been secured, cutting the need for government to respond with emergency food aid. More than 2.5 million people in four of China's poorest provinces—Shanxi, Shaanxi, Gansu,

---

[181] World Bank. (2007). *Restoring China's Loess Plateau*. http://www.worldbank.org/en/news/feature/2007/03/15/restoring-chinas-loess-plateau, accessed 12 November 2016.

and the Inner Mongolia Autonomous Region—have been lifted out of poverty. Further 'downstream' benefits include dramatic reductions in sedimentation of waterways, reducing inputs to the Yellow River by more than 100 million tons each year and slowing the infilling of dams.

The First Loess Plateau project cost US$252 million, of which US$149 million was contributed by the International Development Association (IDA: part of the World Bank). The Second Loess Plateau project cost US$239 million, with an IDA contribution of US$50 million. These investments are sizeable, but physical and economic transformation of the Loess Plateau demonstrates the scale of linked socio-ecological benefits that can be achieved if appropriate ecosystem-based restoration is undertaken in degrading areas. This forms a basis for sustainable outcomes with multiple wider co-benefits arising from close partnership with government, good policies, technical support and active consultation with and participation of local people.

The Loess Plateau Watershed Rehabilitation Projects' principles have since been widely adopted and replicated throughout China. The World Bank estimates that as many as 20 million people have benefited from uptake of the approach.

Box 3.30 Revitalising the Ethiopian Highlands

Natural forest cover in the South Central Rift Valley Region of Ethiopia declined from 16% to 2.8% between 1972 and 2000. This represents a cleared area of 40,324 hectares (about three times the area of Washington DC), an annual loss of 1,440 hectares and a total loss of 82% of the 1972 forest cover, indicative of trends in the wider region.[182] The cumulative consequences of widespread small-scale agriculture and commercial-scale logging and farming were principal drivers of loss, deforestation not only destroying habitat but also reducing water availability. With increasing poverty allied with population growth in the Highlands, pressures arising from the search for subsistence income had driven a spiral of massive deforestation and decreasing habitat functioning.[183]

Re-vegetation of the Ethiopian Highlands was identified as a key requirement to halt and reverse upland desertification and linked poverty.[184]

---

[182] Dessie, G. and Kleman, J. (2007). Pattern and magnitude of deforestation in the South Central Rift Valley Region of Ethiopia. *Mountain Research and Development*, 27(2), pp. 162–168.

[183] Nyssen, J. (1997). Vegetation and soil erosion in Dega Tembien (Tigray, Ethiopia). *Bulletin du Jardin Botanique National de Belgique/Bulletin van de Nationale Plantentuin van België*, 66, pp. 39–62.

[184] Chadhokar, P. and Abate, S. (1988). Importance of revegetation in soil conservation in Ethiopia. In: Rimwanich, S. (Ed.), *Constraints and Solutions to Application of Conservation Practices*. Bangkok, Thailand, pp. 1203–1213.

Significant amongst projects undertaken to address this problem was the *Forest Rehabilitation through Natural Regeneration in Tigray, Northern-Ethiopia*, funded by the Belgian Government.[185] It involved rehabilitation of forests, and the establishment of 'closed areas' where grazing is forbidden or restricted. Replanting trees and shrubs to stabilise ravines and slopes in the Ethiopian uplands are amongst measures reversing a trend of abandonment of formerly severely eroding and barren areas that was also intensifying floods, droughts and soil loss, and leading to a constant requirement for food aid.

Today, Abrha Weatsbha in Tigray region is unrecognisable from its pre-restoration state in the late 1990s, environmental catastrophe averted through planting of millions of tree and bush seedlings. Wells that had run dry are now naturally recharged, the soil is regenerating, fruit trees grow in the valleys and the formerly eroded hillsides are restored to vegetative cover. 'Regreening' has yielded dramatic and surprisingly rapid results in target areas. This has been at relatively little cost to Ethiopian farming communities, who have worked together to exclude grazing from large areas of the most vulnerable land, to replant trees and undertake water conservation measures, and to adopt agro-ecology methods that combine crops and trees on the same pieces of land.[186] Substantial ecosystem service benefits have been found to result from exclosures and similar interventions.[187] Ethiopia's progress is part of a wider trend already transforming degraded and deforested land across Africa.[188]

A 'systemic solutions' approach to reanimate ecosystems and their multiple services across broad landscapes is also evidenced in Europe. One exemplar is extensive installation of integrated constructed wetlands (ICWs) in the Anne Valley in County Waterford, Ireland (Box 3.31). Here, extensive regeneration of lost wetland ecosystem functioning, using both restored and constructed systems, has significantly reanimated not merely the ecology but also the landscape aesthetics and profitability of the farming landscape.

---

[185] *Forest Rehabilitation through Natural Regeneration in Tigray, Northern-Ethiopia* (VL.I.R. EI-2000/PRV-06), September 2000 to September 2004, Belgian Government Vlaamse Interuniversitaire Raad (VL.I.R.) Fund.

[186] Reij, C., World Resources Institute, quoted in: Vidal, J. (2014). Regreening program to restore one-sixth of Ethiopia's land. *The Guardian*, Thursday 30 October 2014. http://www.theguardian.com/environment/2014/oct/30/regreening-program-to-restore-land-across-one-sixth-of-ethiopia, accessed 8 April 2015.

[187] Mekuria, W., Veldkamp, E., Tilahun, M. and Olschewski, R. (2014). Economic valuation of land restoration: The case of exclosures established on communal grazing lands in Tigray, Ethiopia. *Land Degradation and Development*, 22(3), pp. 334–344.

[188] Vidal, J. (2014). Regreening program to restore one-sixth of Ethiopia's land. *The Guardian*, Thursday 30 October 2014. http://www.theguardian.com/environment/2014/oct/30/regreening-program-to-restore-land-across-one-sixth-of-ethiopia, accessed 8 April 2015.

> **Box 3.31 Integrated Constructed Wetlands (ICWs) in County Waterford, Ireland**
>
> The extensive installation of integrated constructed wetlands (ICWs) in the Anne Valley in County Waterford, Ireland, is evidence of a 'systemic solution' using natural processes to achieve multiple ecosystem service benefits. Up until the early 1980s, Waterford was naturally water-rich. Wetlands characterised the landscape, performing a range of hydrological, chemical and biological functions. However, in the 1980s, agriculture improvement subsidies from both the Irish government and the European Union drove the drainage of substantial areas of bog and other wetlands across Ireland. Although land drainage has boosted some facets of agricultural production, drainage of these wetlands produced a number of unintended negative effects on local ecosystems services. These included reducing water storage and floodwater buffering capacity, substantially diminishing rivers and wetlands, altering chemical cycling and degrading biodiversity.
>
> The ICW concept proactively addresses multiple ecosystem service outcomes associated with wetland processes. It takes a 'landscape fit' approach, reinstating cascades of shallow, vegetated wetland cells within natural, aesthetic and working landscapes. Linked benefits include wastewater processing, hydrological buffering, regeneration of flows in watercourses, public access to attractive regenerated wetland landscapes, silt and nutrient interception, and the recovery of lost landscapes and populations of aquatic species such as otters, brown trout, salmon, sea trout and eels. Networks of ICWs in the Anne Valley now support farm profitability, manage sewage from household and industrial units up to village scale, provide leisure opportunities and regenerate the ecology, recreational and aesthetic value of a formerly much degraded catchment ecosystem. Widespread uptake of ICWs has reanimated the Anne Valley, ecologically, socially and economically, with extensive scientific verification of ecosystem service outcomes (reviewed in greater detail in my 2013 book *The Hydropolitics of Dams*[189]).

ICWs have been adopted elsewhere in Ireland for a variety of reasons. These include the treatment of landfill leachate, hotel wastewater and diffuse inputs in a city centre context, with many ecosystem service co-benefits.[190] Regulatory agencies initially resisted granting consents for installation of ICWs, due largely to the narrow terms under which these licences are issued and the exclusion from consideration of the wider suite of benefits they deliver. However, ICW design was subsequently incorpo-

---

[189] Everard, M. (2013). *The Hydropolitics of Dams: Engineering or Ecosystems?* Zed Books, London.
[190] Everard, M., Harrington, R. and McInnes, R.J. (2012). Facilitating implementation of landscape-scale integrated water management: The integrated constructed wetland concept. *Ecosystem Services*, 2, pp. 27–37.

Fig. 3.11 A mature integrated constructed wetland (ICW) on the Ingol catchment in north Norfolk, eastern England, reduces nutrients and improves other aspects of water quality, also buffering flows and providing habitat for wildlife and recreational opportunities amongst a diversity of other valuable ecosystem services. (Image © Rob McInnes)

rated into Irish Government guidance under the *Water Services Investment Programme 2010–2012*,[191] recognising the potential of ICWs to reverse former declines in the ecosystem services of lost natural wetlands.

Whilst not a panacea, ICWs represent low-input, multi-service output 'systemic solutions', averting unintended negative impacts inherent in alternative, single-focus electromechanical treatment technologies and of greater cumulative economic value. There are some pilot implementations in the UK, for example in the Ingol catchment in north Norfolk (eastern England) (Fig. 3.11). However, there is at present resistance to their wider application remains due to the narrow disciplinary focus of many regulatory bodies, an example of 'regulatory lag' as the legislative environment takes often considerable time to evolve with emerging understanding and priorities.

---

[191] Department of the Environment, Heritage and Local Government. (2010). *Integrated Constructed Wetlands: Guidance Document for Farmyard Soiled Water and Domestic Wastewater Applications.* 121pp. http://www.environ.ie/en/Environment/Water/WaterQuality/News/MainBody,24926,en.htm, accessed 3 June 2015.

Further landscape-scale efforts to regenerate ecosystems and the socio-economic welfare that they support include the African Forest Landscape Restoration (AFR100) programme (Box 3.32) and Malawi's ambitious pledge to tackle environmental degradation (3.33).

Box 3.32 African Forest Landscape Restoration (AFR100)

The African Forest Landscape Restoration Initiative (AFR100)[192] aims to restore 100 million hectares of deforested and degraded land across Africa by 2030. AFR100 was launched in December 2015 during the Global Landscapes Forum, part of the CoP (Conference of Parties) of the UNFCCC climate change meetings in Paris. The initiative connects political partners in participating African nations with technical and financial support. The aim is to scale up restoration on the ground to achieve a range of societal benefits including food security, climate change resilience and poverty alleviation. AFR100 provides a platform for more effectively working together to accelerate restoration successes, with activities driven and owned by partner countries.

Financial and technical contributions by international partners support national efforts and deploy resources to partner countries. AFR100 contributes to the Bonn Challenge, the African Resilient Landscapes Initiative (ARLI),[193] goals in the African Union Agenda 2063 (particularly under the Biodiversity and Land Degradation (LDBA) program), the UN Convention on Combating Desertification (UNCCD) and the UN Sustainable Development Goals (SDGs), amongst other targets.

As of June 2019, 27 countries have committed to restoration of land totalling 111 million hectares, including: Benin; Burkina Faso; Burundi; Cameroon; the Central African Republic; Chad; Côte d'Ivoire; the Democratic Republic of the Congo; Ethiopia; Ghana; Guinea; Kenya; Liberia; Madagascar; Malawi; Mozambique; Niger; Nigeria; the Republic of the Congo; the Republic of Sudan; Rwanda; Senegal; South Africa; Swaziland; Tanzania; Togo; and Uganda. An AFR100 Secretariat serves as the main communication hub for the initiative, housed under the NEPAD (New Partnership for Africa's Development) Agency, an economic development program of the African Union.

---

[192] http://www.afr100.org.

[193] World Bank. (2015). *NEPAD Launches Initiative for the Resilience and Restoration of African Landscapes.* The World Bank, Washington, DC. https://www.worldbank.org/en/news/press-release/2015/12/06/nepad-launches-initiative-for-the-resilience-and-restoration-of-african-landscapes.

Box 3.33 Malawi's Ambitious Pledge to Tackle Environmental Degradation

In 2016, the Republic of Malawi made an ambitious 4.5 million hectare restoration pledge under the Bonn Challenge and the African Forest Landscape Restoration Initiative (AFR100).[194] This approach is based on deploying nature-based solutions, such as Forest Landscape Restoration (FLR), to offer an integrated approach to tackling environmental degradation and to enhancing linked human wellbeing. Large-scale landscape restoration is intended to create significant social, economic, and environmental benefits in Malawi, including increased food, water, and livelihood security for many Malawians, through restoration of degraded and deforested land. Malawi's National Forest Landscape Restoration Strategy[195] cited studies finding that 29 metric tons of soil per hectare were being lost annually, reducing productivity of croplands, with the resultant soil erosion and siltation hindering expansion of hydropower generation. Water supplies had been drying up under longer dry seasons, with damage to the water sector estimated at US$11.8 million. Almost 98% of Malawi's cultivated land is rain-fed, with forest loss increasing the vulnerability of farmers to climate change and shifting weather patterns. There was greater vulnerability for agriculture to El Niño effects, and land degradation was costing Malawi an estimated $244 million (6.8% of GDP) per year. The commitment to landscape restoration took the shape of commitment to a 'whole of government' approach, requiring close coordination among different government departments and agencies but also, critically, the involvement of local communities and particularly women who play significant roles in managing lands.

The Republic of Malawi is applying a Restoration Opportunities Assessment Methodology (ROAM) across all districts as the basis for a roadmap for restoration action, identifying areas to be prioritised for action and the most effective interventions to be put in place. Importantly, the process is stakeholder-driven, ensuring that the design and implementation of restoration is firmly rooted in the needs of local communities. The ROAM process has identified 7.8 million hectares available for restoration, the focus shifting now to embedding restoration activities into linked development strategies. This also requires growing the capacity of government agencies

---

[194] Ministry of Natural Resources, Energy and Mining—Malawi. (2017). *Forest Landscape Restoration Opportunities Assessment for Malawi*. NFLRA (Malawi), IUCN, WRI, xv + 126pp.

[195] Ministry of Natural Resources, Energy and Mining—Malawi. (2016). *National Forest Landscape Restoration Strategy*. NFLRA (Malawi), IUCN, WRI, xv + 126pp.

and local communities to conduct FLR interventions, put in place robust monitoring systems and unlock funds. A particular focus is farmer-managed natural regeneration, wherein farmers retain trees on their croplands instead of clearing them. This approach to natural regeneration is already occurring in many places throughout Malawi, as farmers learn that this nature-based approach increases the productivity of their lands.

Some ostensibly nature conservation schemes have also proven to be multi-beneficial in outcome, the conservation goal constituting an 'anchor service' for which systemic solutions yield wider societally beneficial outcomes. A major initiative from the USA is the restoration of the Everglades system in Florida (Box 3.34). Not only has this large-scale restoration reversed the former precipitous decline in the Everglades ecosystem, it has also co-delivered a host of societal benefits of substantial cumulative value.

Box 3.34 Restoration of the Everglades

The Everglades in the US state of Florida once constituted a vast, free-flowing river of grass from the Kissimmee chain of lakes to Florida Bay. This system was rich in wildlife, ranging from flocks of wading and migratory birds, plants and fish, panthers, manatees and deer, and conveyed clean water into Florida Bay and the coral reefs. Development impacting the Everglades was initiated by drainage canals dug since the 1800s. Development accelerated throughout the twentieth century as measures were engineered to provide flood protection and fresh water for the burgeoning population of south Florida. However, it also inflicted substantial harm upon the Everglades ecosystem. Today, more than half the Everglades wetlands have been lost to development and 1,700 miles of canals and levees have vastly changed the landscape[196] and reduced populations of, or entirely extirpated, native birds and other wildlife.

The *Comprehensive Everglades Restoration Program*[197] was initiated in 2000 by US President Bill Clinton. Its longer-term aims included restoring degraded swampland, addressing water quality problems through natural purification, boosting biodiversity including vulnerable species, adding

---

[196] Everglades Restoration. (n.d.). *Restoring America's Everglades.* http://www.evergladesrestoration.gov/, accessed 8 April 2015.

[197] USGS. (2006). *Comprehensive Everglades Restoration Plan (CERP).* http://fl.water.usgs.gov/CERP/cerp.html, accessed 8 April 2015.

value to the tourism industry and restoring natural floodwater controls, amongst a range of other ecosystem services. The Comprehensive Everglades Restoration Program has since become one of the world's largest natural capital restoration projects, generating substantial societal benefits through restored ecosystem functioning.

At an even broader geographical scale, spanning thousands of kilometres, a mosaic of partners and more than 20 countries from the Sahelo-Saharan region have embarked on a bold and visionary *Great Green Wall for the Sahara and Sahel Initiative* (Box 3.35).

Box 3.35 Africa's Great Green Wall

The *Great Green Wall for the Sahara and Sahel Initiative*[198] is an African partnership to tackle desertification and land degradation, affecting millions of the most vulnerable people in Africa in regions where two-thirds of the land cover comprises drylands and deserts. Here, human pressures on fragile ecosystems, compounded by deforestation and climate change, have led to landscape degradation with serious negative impacts on the food security and livelihoods of local communities.

The Great Green Wall initiative focuses on restoration of a 'green' belt running east-to-west across Sahelian Africa some 15 km (9 miles) wide and 7,100 km (4,400 miles) long from Dakar in Senegal to the west, though Mauritania, Mali, Burkina Faso, Niger, Nigeria, Chad, Sudan, Eritrea and to Djibouti in Ethiopia to the east. This swathe of dryland is home to many of the world's poorest and most vulnerable people, the Great Green Wall scheme seeking to restore ecosystem health, functioning and socio-economically valuable services to tackle poverty and soil degradation.[199]

Initially articulated by the President of Nigeria and strongly supported by the President of Senegal in 2005, the vision has grown into an integrated ecosystem management approach adopted by the African Union.[200] Funding has been agreed by the Global Environment Fund[201] to link national-level

---

[198] African Union Commission. (n.d.). *Great Green Wall for the Sahara and the Sahel Initiative*. African Union Commission. http://www.fao.org/docrep/016/ap603e/ap603e.pdf, accessed 10 January 2019.

[199] Global Environment Facility. (2013). *The Great Green Wall*. Global Environment Facility, 11 November 2013. https://www.thegef.org/news/great-green-wall, accessed 10 January 2019.

[200] Global Environment Facility. (n.d.). *The Great Green Wall Initiative*. Global Environment Facility. https://www.thegef.org/gef/great-green-wall, accessed 28 May 2016.

[201] Global Environment Fund. (n.d.). *About GEF*. Global Environment Fund. www.globalenvironmentfund.com/, accessed 28 May 2016.

efforts across borders, improving prospects for overcoming policy, invest-ment and institutional barriers, and averting risks of conflict. In 2007, African Heads of State and Government endorsed the *Great Green Wall for the Sahara and the Sahel Initiative* with the objective of tackling the detri-mental social, economic and environmental impacts of land degradation and desertification in the region. The first practical step towards the Great Green Wall was set with the development of a harmonized strategy under a €1.75 million African Union project launched in September 2011. The ini-tiative supports the efforts of local communities in the sustainable manage-ment and use of forests, rangelands and other natural resources in drylands, as well as contributing to climate change mitigation and adaptation.

From the initial idea of a line of trees from east to west through the African desert, the vision for a Great Green Wall has evolved into that of a mosaic of integrated interventions addressing the challenges facing the people in the Sahel and Sahara. The overall goal of the Great Green Wall initiative is to strengthen the resilience of the region's people and natural systems through sound ecosystem management and the sustainable devel-opment of land resources, underpinning the protection of rural heritage and improvement of living conditions for local populations.

By August 2018, the Great Green Wall was no longer envisioned as a nar-row band of trees along the southern edge of the Sahara, but as a belt of trees and bushes surrounding the Sahara. This wider vision engages all the countries surrounding the Sahara, including Algeria and others in North Africa, in addition to the 11 original sub-Saharan countries of the Sahel. Today, the Initiative has 21 African countries participating, some $8 billion of pledged funding, and weighty partners including the World Bank and the French government. Its ambitions are equally huge: to restore 100 mil-lion hectares of land; provide food security for 20 million people; create 350,000 jobs; and sequester 250 million tons of carbon by 2030.[202] 15% of drought-resistant trees have been planted, largely in Senegal, with 4 million hectares of land restored, and successful grassroots greening efforts in Niger have helped close the gap between the project's ambition and reality.

3.5   Restoring Coastal Functioning

Flat coastal and estuarine lands offer favourable conditions for industrial, port, recreational and residential development as well as conversion for intensive agriculture (Fig. 3.12). Consequently, coastal development is intensive and extensive. However, this brings with it vulnerabilities for

---

[202] Bilski, A. (2018). Africa's Great Green Wall: A work in progress. *Landscape News*, 3 August 2018. https://news.globallandscapesforum.org/28687/africas-great-green-wall-a-work-in-progress/.

constructed and farmed areas and associated built infrastructure in areas prone to flooding. These pressures are compounded by substantial conversion of natural coastal habitats, degrading their many flood and storm regulation, wildlife, fish nursery and other ecosystem services.

This conflict of development with degraded coastal ecosystem processes is brought into sharp focus as coastal defence structures across Europe reach an end to their useful life and integrity, with associated risk of failure. These extensive coastal 'defences' were built under agricultural intensification policies following the Second World War, leading to extensive land drainage and 'reclamation'. to increase food security. Many coastal defence structures are too costly to renew relative to likely farmed benefits, and the politics of public funds supporting private profit-taking from farming activities also make subsidised renewal unlikely. As a consequence, 'managed realignment'—controlled re-flooding as outer defensive walls are breached with construction of smaller set-back defences—is increasingly being adopted to regenerate former intertidal habitat, along with its associated ecosystem services that may also be optimised by foresighted planning (Box 3.36).

Box 3.36 Managed Realignment for Coastal Flood Regulation and Other Benefits

Managed realignment, sometimes called managed retreat, involves the deliberate breaching of artificial seawalls to allow the reinstatement of tidal inundation over areas of formerly 'reclaimed' low-lying land, often converted in past decades or centuries for agricultural purposes. Maintenance and heightening of seawalls are becoming increasingly cost-ineffective, and there is a growing need for habitat restoration enforced across Europe under the EU Habitats Directive. Managed realignment is thus becoming an increasingly common approach. Effectively, managed realignment allows natural regeneration of intertidal saltmarsh and mudflat habitat for their energy dissipation, flood regulation and other beneficial ecosystem services, which can also include habitat for wildlife and nursery areas for juvenile fish.

Around five years after initial breaches in the seawalls, wintering shorebird communities within managed realignment sites were found to be broadly similar to those of surrounding mudflats,[203] and halophytic (salt-

---

[203] Atkinson, P.W., Crooks, S., Drewitt, A., Grant, A., Rehfisch, M.M., Sharpe, J. and Tyas, C.J. (2014). Managed realignment in the UK—The first 5 years of colonization by birds. *Ibis*, 146, pp. 101–110.

tolerant) plant species colonized realignment sites rapidly though specific plant assemblages often differed from more natural sites.[204]

An ecosystem services assessment of the managed realignment at Alkborough Flats on the Humber in north eastern England found that outcomes of the scheme were substantially beneficial, yielding a conservative benefit-to-cost ratio of 3.22 for services that could be quantified.[205] Impacts on provisioning services were largely neutral as grazing with rare sheep breeds compensated for lost arable crop production, with an additional unquantified benefit through providing nursery areas for juvenile fish enhancing stocks. There was also substantial enhancement of flood and climate regulation, recreational and tourism enhancement, and provision of substantial new habitat for wildlife.

The 'anchor services' of flood and natural hazard regulation through managed realignment are of substantial value. Co-beneficial ecosystem services can be optimised around them, including habitat for wildlife, nursery areas for fish, nutrient cycling and carbon sequestration, amenity and potential new productive uses including saltmarsh grazing, shellfisheries and harvesting of edible plants such as *Salicornia* (glasswort).[206] Estuaries and intertidal habitats are significant for recruitment of fish species of recreational and commercial importance, optimisation of this service in the design of managed realignment schemes potentially supporting stock recruitment as a better alternative than controlling fishery exploitation.[207,208]

Managed realignment for coastal flood regulation and other linked ecosystem services is just one example of renewed interest in working in synergy with natural coastal processes. Another is recognition of the value of coastal habitats for natural hazard regulation, amongst a range of linked ecosystem services. Sand dunes, for example, provide many ecosystem services, significantly including natural hazard regulation. Historic tendencies to convert sand dunes are being superseded by protection and natural

[204] Mossman, H.L., Davy, A.J. and Grant, A. (2012). Does managed coastal realignment create saltmarshes with 'equivalent biological characteristics' to natural reference sites? *Journal of Applied Ecology*, 49, pp. 1446–1456.

[205] Everard, M. (2009). *Ecosystem Services Case Studies*. Environment Agency, Bristol.

[206] Watts, W.D., Colclough, S. and Everard, M. (2018). Economics of wetland conservation case study: Learning from managed realignment. In: Finlayson, C.M., Everard, M., Irvine, K., McInnes, R.J., Middleton, B.A., van Dam, A.A. and Davidson, N.C. (Eds), *The Wetland Book: I: Structure and Function, Management, and Methods*. Springer, Dordrecht, pp. 917–924.

[207] Everard, M. (2014). Realising sustainable fisheries. *Fish*, 113(Spring 2014), pp. 14–19.

[208] Everard, M. (2014). Regenerating coastal fish populations. *Fish*, 114(Summer 2014), pp. 12–14.

Fig. 3.12 Estuaries are important for many ecosystem services, though many are now extensively developed for industry, ports, habitation or agriculture, offering substantial potential for regeneration of ecosystem processes yielding a diversity of linked societal benefits. (Image © Dr Mark Everard)

regeneration recognising their significant, if formerly overlooked, values including as diverse habitat matrices for wildlife, amenity, tourism, military and other uses.[209] It is hoped and anticipated that growing awareness of these tangible benefits can stem the tide in degradation of these important transitional habitats. Coastal forests (in particular mangroves), coral reefs, seagrasses, kelp forests, wetlands and dunes are observed to be capable of mitigating the effects of natural hazards such as coastal storms,[210]

[209] Everard, M., Jones, L. and Watts, W. (2010). Have we neglected the societal importance of sand dunes? An ecosystem services perspective. *Aquatic Conservation: Marine and Freshwater Ecosystems*, 20, pp. 476–487.

[210] Wells, S., Ravilious, C., and Corcoran, E. (2006). *In the Front Line: Shoreline Protection and Other Ecosystem Services from Mangroves and Coral Reefs*. UNEP World Conservation Monitoring Centre, Cambridge, UK.

hurricanes[211] and tsunamis.[212] A diverse range of species also interact in coastal environments to bind and stabilize sediments creating natural sea defences, for example salt marshes, mangrove forests, kelp forests and sea grass beds.[213]

Contemporary 'hard' protective coastal and flood defences can ironically increase the flood vulnerability of built infrastructure as land margins shrink and remobilise carbon content when deprived of sediment inputs, so can no longer 'grow with the sea'.[214] Many of the world's major deltas—among them the Nile, Brahmaputra-Ganges, Yangtze, Irrawaddy and Mekong—suffer compounding pressures of starved sediment feed due to upstream impoundments trapping suspended matter, sea level rise, and degraded deltaic habitat. Where this occurs, there is significant potential for damaging impacts, particularly where intensified by storminess and tsunamis, vulnerabilities exacerbated where natural coastal defensive habitats and processes are most degraded.[215] The potential costs of this are illustrated by flooding of New Orleans after landfall of Hurricane Katrina in 2005, together with coastal proposals for a nature-based approach to restoration (Box 3.37). The strategic solution is to stop fighting against but instead to work with natural processes rebuilding land (Fig. 3.13), albeit that inherently dynamic geomorphological (habitat-forming and erosion) coastal processes conflict with narrowly spatially defined land ownership.

---

[211] Costanza, R. Graumlich, L., Steffen, W., Crumley, C., Dearing, J., Hibbard, K., Leemans, R., Redman, C. and Schimel, D. (2007). Sustainability or collapse: What can we learn from integrating the history of humans and the rest of nature. *Ambio*, 36(7), pp. 522–527.

[212] Kathiresan, K. and Rajendran, N. (2005). Coastal mangrove forests mitigated tsunami. *Estuarine, Coastal and Shelf Science*, 65, pp. 601–606.

[213] Rönnbäck, P., Kautsky, N., Pihl, L., Troell, M., Söderqvist, T. and Wennhage, H. (2007). Ecosystem goods and services from Swedish coastal habitats: Identification, valuation, and implications of ecosystem shifts. *Ambio*, 36, pp. 534–544.

[214] Reise, K. (2013). *A Natural History of the Wadden Sea: Riddled by Contingencies*. Wadden Academy, Leeuwarden, The Netherlands. http://www.waddenacademie.nl/fileadmin/inhoud/pdf/04-bibliotheek/13-031_SYN_WADDENLEZING_KReise_lowres.pdf, accessed 25 June 2016.

[215] Everard, M., Kangabam, R., Tiwari, M.K., McInnes, R.J., Kumar, R., Talukdar, G.H., Dixon, H., Joshi, P., Allan, R., Joshi, D. and Das, L. (2019). Ecosystem service assessment of selected wetlands of Kolkata and the Indian Gangetic Delta: Multi-beneficial systems under differentiated management stress. *Wetlands Ecology and Management*, 27(2), pp. 405–426. https://doi.org/10.1007/s11273-019-09668-1.

Box 3.37 2005 Flooding of New Orleans on Landfall of Hurricane Katrina

Landfall of Hurricane Katrina in Louisiana, USA, in August 2005 resulted in the largest natural disaster in US history. 80% of New Orleans was flooded by up to 4.6 m of water[216] as the storm surge breached levees and flood-walls, the weaknesses in which were well known.[217] Total damage from Hurricane Katrina is estimated at $108 billion at 2005 prices,[218] with a final death toll of 1,464 people.[219]

New Orleans was founded on high ground along the Mississippi River but sprawled and extended with navigable waterways, lowering the water table with consequent land shrinkage. Extensive farming in the upstream floodplains of the Mississippi River, walled off from the river channel by levees (flood banks), deprived the river of its substantial input of sediment, further reducing the replenishment of habitat in the lower river. Flood walls around New Orleans compounded starvation of sediment inputs from the already depleted river, the city and environs subsiding into a bowl below sea level. This led scientists to declare *"New Orleans is a disaster waiting to happen"* following near failure of defences during Hurricane Georges in 1998.[220]

In the outer delta, natural coastal protection declined as marshes were no longer nourished by sediment inputs from the Mississippi. Further drainage and construction activities combined to degrade coastal wetlands at an average rate of 65 km² (21 miles²) per year in the first six years of the twenty-first century, with almost 5,000 km² (1800 miles²) of coastal wet-lands lost since the 1930s.[221] Intact wetlands would have averted much of the damage from Hurricane Katrina.

Conventional rebuilding of the city and its defences cannot avert future 'natural disasters'. Wetland restoration through ecological engineering rep-

---

[216] Murphy, V. (2005). Fixing New Orleans' thin grey line. *BBC News*, 4 October 2005. http://news.bbc.co.uk/1/hi/world/americas/4307972.stm, accessed 25 May 2016.

[217] McQuaid, J., Walsh, W., Barnett, J. and Schleifstein, M. (2005). Levees' weakness well-known before breaches: Lack of political will, funds cited in failure. *The Times-Picayune*, 2 September 2005. http://levees.org/wp-content/uploads/2010/07/levee.pdf, accessed 25 May 2016.

[218] Knabb, R.D., Rhome, J.R. and Brown, D.P. (2005). *Hurricane Katrina: August 23–30, 2005 (PDF) (Tropical Cyclone Report)*. National Hurricane Center, United States National Oceanic and Atmospheric Administration's National Weather Service. http://www.nhc.noaa.gov/data/tcr/AL122005_Katrina.pdf, accessed 25 May 2016.

[219] Boyd, E. (2006). *Preliminary Summary Report of Hurricane Katrina Deceased Victim Recovery Locations in Louisiana*. LSU Hurricane Public Health Center.

[220] Fischetti, M. (2001). New Orleans is a disaster waiting to happen. *Scientific American*, October.

[221] Costanza, R., Day, J.W. and Mitsch, W.J. (2006). A new vision for New Orleans and the Mississippi delta: Applying ecological economics and ecological engineering. *Frontiers in Ecology and Environment*, 4(9), pp. 465–472.

resents a cheaper, more sustainable and multi-beneficial solution. Coastal wetlands in Louisiana have been estimated to provide $940 per hectare per year in storm and flood protection services,[222] with additional ecosystem services increasing their net value to $12,700 per hectare per year.[223]

The value of protecting various areas of the sea for the many benefits they provide has been increasingly recognised. Various forms of marine protected area (MPA) have been implemented around the world to safeguard biodiversity, including maintaining fish stocks, amongst a range of other ecosystem services. This applies as much to the open sea as to coastal areas as, though seemingly uniform in appearance, the marine environment is in reality a complex mosaic of areas of differing depths, upwellings, currents, plankton production, and breeding and nursery areas (Fig. 3.14). There is wide acknowledgement of the contribution of protected, non-harvested marine areas important for fish reproduction and nurseries enhancing fish stocks beyond the boundaries of the protected areas.[224] MPAs have been found to increase fish biomass,[225] albeit that outcomes are variable and dependent to a significant extent on the specific features of the MPA.[226] Consistent with agreements by member nations under the Convention on Biological Diversity, the number of MPAs across the world is increasing. However, MPAs are hard to police in practice, for example preventing illegal fishing or the activities of nations still permitting damaging harvesting methods. A review of nearly 1,000 sites in 87 MPAs distributed across the world assessed five key features of known importance: (1) degree of fishing permitted within

---

[222] Costanza, R., Farber, S.C. and Maxwell, J. (1989). The valuation and management of wetland ecosystems. *Ecological Economics*, 1, pp. 335–361.

[223] Costanza, R., d'Arge, R., de Groot, R., et al. (1997). The value of the world's ecosystem services and natural capital. *Nature*, 387, p. 253.

[224] Gell, F.R. and Roberts, C.M. (2003). Benefits beyond boundaries: The fishery effects of marine reserves. *Trends in Ecology and Evolution*, 18, pp. 448–455.

[225] Worm, B. (2017). How to heal an ocean. *Nature*, 543, pp. 630–631.

[226] Claudet, J., Osenberg, C.W., Benedetti-Cecchi, L., Domenici, P., García-Charton, J.A., Pérez-Ruzafa, A., Badalamenti, F., Bayle-Sempere, J., Brito, A., Bulleri, F., Culioli, J.M., Dimech, M., Falcón, J.M., Guala, I., Milazzo, M., Sánchez-Meca, J., Somerfield, P.J., Stobart, B., Vandeperre, F., Valle, C. and Planes, S. (2008). Marine reserves: Size and age do matter. *Ecological Letters*, 11, pp. 481–489.

Fig. 3.13 Coastal wetlands, such as tropical mangrove systems, can play major roles in flood and storm buffering as well as supporting nurseries for marine fisheries, carbon storage, nutrient cycling, microclimate regulation and many more linked benefits. (Image © Dr Mark Everard)

MPAs; (2) level of enforcement; (3) MPA age; (4) MPA size; and (5) presence of continuous habitat allowing unconstrained movement of fish across MPA boundaries.[227] This comparative study considered these factors simultaneously, finding that MPAs that work effectively include at least four of the five key features.

There is growing interest, albeit with many remaining obstacles, to more widespread and strategically located implementation of MPAs.

[227] Edgar, G.J., Stuart-Smith, R.D., Willis, T.J., Kininmonth, S., Baker, S.C., Banks, S., Barrett, N.S., Becerro, M.A., Bernard, A.T.F., Berkhout, J., uxton, C.D., Campbell, S.J., Cooper, A.T., Davey, M., Edgar, S.C., Försterra, G., Galván, D.E., Irigoyen, A J., Kushner, D.J., Moura, R., Parnell, P.E., Shears, N.T., Soler, G., Strain, E.M.A. and Thomson, R.J. (2014). Global conservation outcomes depend on marine protected areas with five key features. *Nature*, 506, pp. 216–220.

Fig. 3.14 Though often appearing uniform from the surface, the marine environment is a complex mosaic of depths, upwellings, currents, plankton production, fishery recruitment and other zones. (Image © Dr Mark Everard)

Such a network, established in areas important for a range of natural processes, could create 'wilder' zones of the marine environment to protect or allow restoration of their inherent natural value, but also to safeguard natural processes supporting ecosystem regeneration and the continued flow of benefits to humanity. Despite a global agreement to protect 10% of the world's oceans by 2020, only about 7% are currently under some form of protection, with significantly less under effective management.[228] However, the economic returns from effective ocean management can potentially create billions of dollars of benefits.[229]

[228] Gill, D.A., Mascia, M.B., Ahmadia, G.N., Glew, L., Lester, S.E., Barnes, M., Craigie, I., Darling, E.S., Free, C.M., Geldman, J., Holst, S., Jensen, O.P., White, A.T., Basurto, X., Coad, L., Gates, R.D., Guannel, G., Mumby, P.J., Thomas, H., Whitmee, S., Woodley, S. and Fox, H.E. (2017). Capacity shortfalls hinder the performance of marine protected areas globally. *Nature*, 543, pp. 665–669.

[229] Reuchlin-Hugenholtz, E. and McKenzie, E. (2015). *Marine Protected Areas: Smart Investments in Ocean Health*. WWF, Gland, Switzerland.

3.6   Managing the Uplands

The world's mountains support rich and characteristic biodiversity and geodiversity. They also play significant roles in regulating the global climate system, formation and conservation of soil, capture and storage of water resources, hosting often unique wildlife including serving roles in bird flyways, and providing natural beauty and recreational opportunities, wilderness and distinctive cultural diversity.[230] They thereby provide ecosystem services supporting livelihoods and economic opportunities for a large proportion of the human population, both within the mountain region as well as downstream through catchment systems serving surrounding lowlands.[231,232,233] However, mountains are amongst the most fragile of global ecosystems due to their high sensitivity to a variety of natural and anthropogenic factors. These include high relief, steepness of terrain, often high and erosive rainfall, shallow soils commonly on unstable geologic formations, sensitive and often endemic flora and fauna, climatic complexities and geodynamic instability.[234] Himalayan mountain ecosystems, the world's most extensive, are additionally the world's most densely populated.[235] Climate change compounds stresses

---

[230] Alfthan, B., Gupta, N., Gjerdi, H.L., Schoolmeester, T., Andresen, M., Jurek, M. and Agrawal, N.K. (2018). *Outlook on Climate Change Adaptation in the Hindu Kush Himalaya*. Mountain Adaptation Outlook Series. United Nations Environment Programme, GRID-Arendal and the International Centre for Integrated Mountain Development, Vienna, Arendal and Kathmandu. www.unep.org, www.grida.no, www.icimod.org.

[231] Messerli, B. and Ives, J.D. (Eds.). (1997). *Mountains of the World—A Global Priority. A Contribution to Chapter 13 of Agenda 21*. Parthenon, New York.

[232] Körner, K., Ohsawa, M. and Spehn, E. (2005). Mountain systems. In: Hassan, R., Scholes, R. and Ash, N. (Eds.), *Ecosystems and Human Wellbeing. Current State and Trends: Findings of the Condition and Trends Working Group*. Millennium Ecosystem Assessment Vol 1. Island Press, Washington, DC, Chap. 24, pp. 681–716.

[233] Sharma, E., Chettri, N. and Oli, K.P. (2010). Mountain biodiversity conservation and management: A paradigm shift in policies and practices in the Hindu Kush-Himalayas. *International Journal of Ecological Research*, 25, pp. 909–923.

[234] ICIMOD. (2010). *Mountains of the World: Ecosystem Services in a Time of Global and Climate Change: Seizing Opportunities—Meeting Challenges*. Framework paper prepared for the Mountain Initiative of the Government of Nepal by ICIMOD and the Government of Nepal, Ministry of Environment.

[235] Tiwari, P.C. and Joshi, B. (2015). Local and regional institutions and environmental governance in Hindu Kush Himalaya. *Environmental Science and Policy*, 49, pp. 66–74.

in mountain environments through increasing mean annual temperatures and the melting of glaciers and snow, altered precipitation patterns, and more frequent and extreme weather events.[236] These processes cumulatively intensify impacts on other natural as well as socio-economic drivers of change. These can have significant impacts on flows of ecosystem services, including the water, health, food and livelihood security of people both within mountains and in dependent lowlands.

The high cultural diversity found amongst indigenous mountain communities also embodies a rich resource of locally adapted traditional knowledge, resource development and environment conservation practices, agricultural and food systems, adaptation and coping mechanism, languages, customs, traditions, costumes, conventions and rituals.[237] These have practical significance for environmental resilience and restoration, climate change adaptation and ensuring the sustained resource productivity of mountain ecosystems.[238] This resource of traditional knowledge represents a substantial and often overlooked repository highly germane to the conservation, management and governance of these and other natural resources.[239] Mountain communities have thereby contributed significantly to the conservation and protection of ecosystem goods and services.[240]

Even in relatively low topography nations and islands, such as the British Isles, upland areas including mountains and higher-altitude moorlands and heaths comprise significant areas of semi-natural habitats and landscapes.

[236] Tse-ring, K., Sharma, E., Chettri, N. and Shrestha, A. (2009). *Climate Change Impact and Vulnerability in the Eastern Himalayas—Synthesis Report.* ICIMOD, Kathmandu.

[237] Körner, K., Ohsawa, M. and Spehn, E. (2005). Mountain systems. In: Hassan, R., Scholes, R. and Ash, N. (Eds.), *Ecosystems and Human Wellbeing. Current State and Trends: Findings of the Condition and Trends Working Group.* Millennium Ecosystem Assessment Vol. 1. Island Press, Washington, DC, Chap. 24, pp. 681–716.

[238] ICIMOD. (2010). *Mountains of the World: Ecosystem Services in a Time of Global and Climate Change: Seizing Opportunities—Meeting Challenges.* Framework paper prepared for the Mountain Initiative of the Government of Nepal by ICIMOD and the Government of Nepal, Ministry of Environment.

[239] UNEP-WCMC. (2002). *Mountain Watch: Environmental Change and Sustainable Development in Mountains.* UNEP, Nairobi. www.unep-wcmc.org/mountains/mountainwatchreport/, accessed 11 March 2012.

[240] Jodha, N.S. (2002). *Rapid Globalisation and Fragile Mountains: Sustainability and Livelihood Security Implications in Himalayas.* Final narrative report of the research planning project submitted to the MacArthur Foundation. ICIMOD, Kathmandu, Nepal.

Though substantially modified in character and area by human pressures, these uplands nonetheless produce significant ecosystem services. They constitute sources and areas for purification of water, buffering of hydrology, carbon storage, support for biodiversity including many rare species, and culturally and recreationally important landscapes.[241]

There is growing awareness and appreciation of the values provided by uplands at both global and national scales, providing resources and security to a substantial proportion of the global human population even very remotely from the mountain regions themselves. Mountains have consequently been considered as 'water towers for humanity'[242] (Box 3.38) (Fig. 3.15).

Box 3.38 Mountains Serving as 'Water Towers for Humanity'

Extensive research has been devoted to the many ecosystem services provided by the Hindu Kush Himalayas (HKH), extending for 3,500 km over all or part of eight countries from Afghanistan in the west to Myanmar in the east. Very significantly, the HKH, including the world's tallest mountains, serve as 'the world's water tower' as they are the source of ten large Asian river systems: the Amu Darya, Indus, Ganges, Brahmaputra (Yarlungtsanpo), Irrawaddy, Salween (Nu), Mekong (Lancang), Yangtze (Jinsha), Yellow River (Huanghe), and Tarim (Dayan). These rivers cumulatively provide water and a wide range of other ecosystem services, supporting the livelihoods of around 240 million people in the region and providing water in downstream basins to 1.9 billion people.[243] These benefits extend right down to river deltas hundreds or thousands of miles away, supporting irrigation and domestic uses of water, fisheries, tourism, transportation and hydropower.[244]

---

[241] Van der Wal, R., Bonn, A., Monteith, D., Reed, M., Blackstock, K., Hanley, N., Thompson, D., Evans, M. and Alonso, I. (2011). *Chapter 5: Mountains, Moorlands and Heaths*. UK National Ecosystem Assessment: Technical Report, pp. 105–160.

[242] Viviroli, D., Dürr, H.H., Messerli, B., Meybeck, M. and Weingartner, R. (2007). Mountains of the world, water towers for humanity: Typology, mapping, and global significance. *Water Resources Research*, 43(7). https://doi.org/10.1029/2006WR005653.

[243] Scott, C.A., Zhang, F., Mukherji, A. (coordinating lead authors), Immerzeel, W., Bharati, L., Mustafa, D. (lead authors), Zhang, H., Albrecht, T., Lutz, A., Nepal, S., Siddiqi, A., Kuemmerle, H., Qadir, M., Bhuchar, S., Prakash, A. and Sinha, R. (contributing authors). (2019). Water in the Hindu Kush Himalaya. In: Wester, P., Mishra, A., Mukherji, A., Shrestha, A. B. (Eds.), *The Hindu Kush Himalaya Assessment—Mountains, Climate Change, Sustainability and People*. Springer, Chap. 8. https://doi.org/10.1007/978-3-319-92288-1_8.

[244] Alfthan, B., Gupta, N., Gjerdi, H.L., Schoolmeester, T., Andresen, M., Jurek, M. and Agrawal, N.K. (2018). *Outlook on Climate Change Adaptation in the Hindu Kush Himalaya*. Mountain Adaptation Outlook Series. United Nations Environment Programme, GRID-Arendal and the International Centre for Integrated Mountain Development, Vienna, Arendal and Kathmandu. www.unep.org, www.grida.no, www.icimod.org.

Ethiopia's highlands are known as 'East Africa's Water Tower' for their functions of intercepting and storing rainfall, and consequently as a source of many major east African river systems.[245]

The Kingdom of Lesotho is known as the 'Kingdom in the Sky', as more than 80% of the country is at or above 800 metres above sea level. The highlands of Lesotho are known as 'southern Africa's water tower, as the sources of many of the major rivers of southern Africa are in this upland region.

Management of mountain environments has traditionally constituted a trans-national matter of global and diplomatic significance.[246] This is due to the fact that, in many parts of the world, particularly including Europe, Asia, South America and Africa, mountains form political boundaries. However, flows of a range of ecosystem services of high significance, including the sources and main stems of major transboundary rivers, do not recognise political borders. A collaborative international approach is therefore required for their sustainable management, including bilateral agreements as well as under international customary law.[247] This also necessitates participation at all levels of society and the innovation of strategies to quantify and address threats that transcend linguistic, worldview, political and socio-economic differences.[248] Box 3.39 outlines examples of cross-border mountain sharing, research and management in Africa.

---

[245] Gebrehiwot, S.G., Gärdenäs, A.I., Bewket, W. and Seibert, J. (2014). The long-term hydrology of East Africa's water tower: Statistical change detection in the watersheds of the Abbay Basin. *Regional Environmental Change*, 14, pp. 321–331.

[246] Price, M.F. (2015). Transnational governance in mountain regions: Progress and prospects. *Environmental Science and Policy*, 49, pp. 95–105.

[247] Food and Agriculture Organization of the United Nations (FAO). (2002). *Law and Sustainable Development since Rio—Legal Trends in Agriculture and Natural Resource Management*. FAO Legal Office, Rome, Italy.

[248] United Nations Environment Programme (UNEP). (2014). *African Mountains Atlas*. UNEP, Nairobi.

Fig. 3.15 Mountain ecosystems such as the Himalayas, as here in Bhutan, can serve as 'water towers', carbon stores, biodiversity reserves, sources of food and medicinal resources, tourism and recreational destinations, and repositories of spiritual and artistic inspiration amongst a wide range of linked ecosystem service benefits. (Image © Dr Mark Everard)

Box 3.39 Cross-Border Mountain Sharing and Management in Africa

Cross-border mountain sharing, research and management in Africa has been increasing in recent decades.[249] Examples include:

- Transboundary management of the Virunga Mountains (the 'Greater Virunga Landscape') on an international basis dates back to the 1920s;[250]
- The Mount Nimba Strict Reserve in mountains between Guinea and the Ivory Coast was established in 1981 spanning 175 km^2;
- A 10-year Transboundary Strategic Plan has been implemented covering the Nyungwe-Kibira ecosystem co-managed by Rwanda and Burundi, covering an area of 1,000 km^2;
- The Maloti-Drakensberg Park spanning 2428 km^2 was established in 2001 between South Africa and Lesotho, subsequently designated as a UNESCO World Heritage Site in 2013;
- The Central Albertine Rift Transfrontier Protected Area Network, spanning the tri-border region of the Democratic Republic of the Congo, Rwanda and Western Uganda, covers the Kibale, Mgahinga Gorilla, Queen Elizabeth, Rwenzori Mountains, Semuliki, Virunga and Volcanoes National Parks and Bwindi Impenetrable Forest. This Network is the subject of ten-year strategic plans building on initiatives since 2001;
- The African Highlands Initiative (AHI) was established in 2006 to address livelihoods in mountain areas of Ethiopia, Kenya, Tanzania, Rwanda and Uganda with high human populations;
- Elements of the East African Community Treaty, covering Burundi, Kenya, Rwanda, Tanzania and Uganda since 2000, pertain to management of tourism, wildlife including conservation policies and wider environmental factors. The Treaty includes a Protocol on Environment and Natural Resources Management, one Article of which specifically focuses on mountain habitats; and
- The Africa-wide New Partnership for Africa's Development (NEPAD) contains mountain-specific elements, and its Action Plan for the Environment supports the acknowledged priority of managing watersheds sustainably across the continent.[251]

---

[249] United Nations Environment Programme (UNEP). (2014). *African Mountains Atlas*. UNEP, Nairobi.

[250] United Nations Environment Programme (UNEP). (2014). *African Mountains Atlas*. UNEP, Nairobi.

[251] New Partnership for Africa's Development (NEPAD). (2003). *Action Plan for the Environment Initiative*. NEPAD, Midrand.

Collaboration around management of mountain regions globally therefore includes some important examples of national and transnational efforts to safeguard ecosystems of vital importance. They are recognised not just for inherent reasons, but also to safeguard dependent human livelihoods both within and at often distant remove from the mountains themselves.

3.7   Putting Species to Work

The preceding schemes recognise the central role of rehabilitation or, in some cases, emulation of natural processes for socio-economic security and opportunity. Whilst all species are interactive elements of the ecosystems of which they are part, some are particularly influential in shaping ecosystem structure and functioning.

Some species are particularly active through a process known as 'trophic cascade', occurring when a trophic level in a food web is suppressed. In so-called 'top down' trophic cascade, predators reduce the abundance or alter the behaviour of prey species. We exploit a similar process when removing crop-raiding, stock-predating or other perceived pests to increase the yield of farmed commodity species. Where populations of predators are suppressed or expunged in wider ecosystems, including for example wolves, big cats and other species seen as posing risks to humans and/or livestock, booming uncontrolled herbivore prey can intensify grazing pressure. Disproportionately high herbivore population can in turn influence the structure and functioning of vegetation and sediment movement, with a cascading effect on the functioning and character of the whole ecosystem. Ecosystem change, for example in the damage inflicted on trees by artificially high deer numbers, can create a consequent need to fence out or cull deer to protect forests and retain landscape character.[252] Conversely, when predators are reintroduced, dramatic changes in ecosystem functioning of potentially significant benefit to

---

[252] Armstrong, H., Gill, R., Mayle, B. and Trout, R. (n.d.). *Protecting Trees from Deer: An Overview of Current Knowledge and Future Work*. Forest Research. https://www.forestresearch.gov.uk/documents/321/FR0102deer_TeTD1Jj.pdf, accessed 3 January 2019.

ecosystem stability and flows of benefits to society may ensure. We see this in the case of the reintroduction of wolves into Yellowstone National Park in the USA (see Box 3.40) and the recovery of sea otter populations on the Pacific coast of America (Box 3.41).

Box 3.40 Ecosystem Functioning Effects of Grey Wolf Reintroductions

Surprising results may ensue when top predators are reintroduced or allowed to recover. Restoration of the grey wolf (*Canis lupus*) into several areas in the northern Rocky Mountains of the United States, including in Yellowstone National Park since 1995, has changed the character and functioning of American wilderness ecosystems.[253]

Benefits from wolf reintroductions in Yellowstone were almost immediate. The wolves controlled numbers, increased health and changed the behaviour of wapiti (a deer: *Cervus canadensis*), which spent less time in valleys and gorges where wolves easily ambush them. This promoted re-establishment of river corridor flora, increasing biodiversity and providing food and shelter for a growing variety of plants and animals. Vegetative proliferation also decreased riverbank erosion, stabilising and deepening channels. 'Renaturalisation' of predatory control generated multiple, interconnected benefits flowing downstream to people.

Box 3.41 Ecosystem Functioning Effects of Sea Otter Recovery

Another dramatic impact observed in the US is the recovery of sea otters (*Enhydra lutris*) on the Pacific coast. Otters were a top predator in this ecosystem, but the species had previously been hunted to near-extinction. The absence of otters enabled a substantial boost in the population of pacific purple sea urchins (*Strongylocentrotus purpuratus*), one of the primary prey species. This in turn substantially increased grazing pressure on giant kelp (*Macrocystis pyrifera*), leading to massive deterioration of the kelp forests along the California coast together with their many energy dissipating, fish nursery, nutrient cycling, habitat and other beneficial functions.

---

[253] The Guardian. (2014). How wolves change rivers—Video. *The Guardian* (GrrlScientist), 3 March 2014. http://www.theguardian.com/science/grrlscientist/2014/mar/03/how-wolves-change-rivers, accessed 8 April 2015.

Subsequent protective measures have enabled sea otter populations to re-establish in Monterey Bay, with restoration of former ecosystem structure, functioning and its many benefits.[254] In 2015, sea otter numbers in Monterey Bay topped 3,000 for the first time since fur traders decimated the population in the nineteenth century. Though good news for the otters, fishing groups had become concerned that sea otters might suppress the rebound in stocks of endangered black abalone (*Haliotis cracherodii*).[255] However, a detailed study investigating abalone populations in twelve places along the Central Pacific Coast of the USA, some with otters and some without, found that more abalone occurred where there were more otters.[256] This is believed to be due to the otters controlling numbers of sea urchins, part of the otters' staple diet, preventing overgrazing of rich kelp forests over rocky bottoms. Otter activity improved the health of kelp forests to the benefit of co-evolved species including abalone.

Other species are referred to as 'ecosystem engineers' as they are particularly influential in shaping the ecosystems of which they are part.[257] A classic example here is the action of beavers in modifying stream structure (Fig. 3.16). Beaver reintroductions in places where they had been previously exterminated have produced many beneficial outcomes (Box 3.42). Similar beneficial outcomes have also resulted from reintroduction of bears (Box 3.43).

[254] U.S. Fish and Wildlife Service. (2015). *Southern Sea Otter* (Enhydra lutris nereis)—*5-Year Review: Summary and Evaluation*. U.S. Fish and Wildlife Service, Ventura Fish and Wildlife Office, Ventura, CA. https://www.fws.gov/ventura/docs/species/sso/Southern%20Sea%20Otter%20 5%20Year%20Review.pdf, accessed 3 January 2019.

[255] Monterey Bay Aquarium. (2015). *Sea Otters and Abalone: A Special Synergy*. Conservation & Science at the Monterey Bay Aquarium. https://futureoftheocean.wordpress.com/2015/12/18/ sea-otters-and-abalone-a-special-synergy/.

[256] Raimondi, P., Jurgens, L.J. and Tinker, M.T. (2015). Evaluating potential conservation conflicts between two listed species: Sea otters and black abalone. *Ecology*, 96(11), pp. 3102–3108.

[257] Jones, C.G., Lawton, J.H. and Shachak, M. (1994). Organisms as ecosystem engineers. *Oikos*, 69(3), pp. 373–386.

Fig. 3.16 A beaver lodge in the Rockies National Park, modifying and diversifying riverine habitat resulting in increased niches, biodiversity and beneficial ecosystem services. (Image © Rob McInnes)

Box 3.42 Beaver Reintroductions

Beavers are 'ecosystem engineers', their damming activities changing the dynamics of river ecosystems. Reintroduction and reestablishment of beavers offers a natural mechanism for restoring degraded streams.

Restoration of North American beavers (*Castor canadensis*) has had a positive influence on biodiversity. They create and diversify habitats at catchment scale by creating ponds and wetlands, altering sediment transport processes, importing woody debris into aquatic environments, and creating important habitat and successional features.[258] Reintroductions of North American beavers were also found to enhance stream habitat beneficially for endangered Chinook and Steelhead salmon in the Pacific Northwest in the Upper Columbia River Basin in Eastern Washington,

---

[258] Law, A., McLean, F. and Willby, N.J. (2016). Habitat engineering by beaver benefits aquatic biodiversity and ecosystem processes in agricultural streams. *Freshwater Biology*, 61(4), pp. 486–499.

USA.[259] A study in the Pacific Northwest recorded a 52% survival increase in juvenile Steelheads, with beaver-related habitat diversification also providing nesting and foraging territory for passerine birds, with improved habitat also utilised by bats.[260]

Reintroduction of Eurasian beavers (*Castor fiber*) into large enclosures in southern England significantly increased water storage and quality, yielding multiple benefits for ecosystems and services provided to people.[261] Beaver reintroductions are not without public concern. Residents in Alyth, Perthshire (Scotland), believed that dams built upstream by renaturalised beavers contributed to flooding of the town.[262] However, subsequent scientific study comparing multiple beaver-modified and unmodified sites on headwater streams in eastern Scotland observed that dam building by beavers made a major contribution to restoring natural habitats and functions in degraded streams, holding back water in the landscape and regulating fluctuations in stream flow as well as increasing organic matter retention, plant biomass and macroinvertebrate diversity at landscape scale.[263]

Box 3.43 Bear Reintroductions

Conservation of America's bear species depends on the health of the forests that support them. Bears are associated with a wide range of linked benefits including carbon storage, storm buffering, regulation of flooding, and the storage and purification of water. Bears play reciprocal roles as predators, agents of seed dispersal and ecosystem engineers, maintaining the diversity, functioning and services of their environments.[264]

---

[259] Methow Beaver Project. (2013, April). *Methow Beaver Project: Accomplishments and Outcomes.* Methow Beaver Project. http://www.methowrestorationcouncil.org/MethowBeaverProjectReport2013. pdf, accessed 8 April 2015.

[260] Bouwes, N., Weber, N., Jordan, C.E., Saunders, W.C., Tattam, I.A., Volk, C., Wheaton, J.M. and Pollock, M.M. (2016). Ecosystem experiment reveals benefits of natural and simulated beaver dams to a threatened population of steelhead (*Oncorhynchus mykiss*). *Scientific Reports*, 6, p. 28581. https://doi.org/10.1038/srep28581.

[261] Devon Wildlife Trust. (n.d.). *Devon Beaver Project.* http://www.devonwildlifetrust.org/devon-beaver-project/, accessed 18 May 2015.

[262] Jackson, R. (2015). Residents blame Beavers for flash floods in Alyth. *The Scotsman*, 22 July 2015. https://www.scotsman.com/news/environment/residents-blame-beavers-for-flash-floods-in-alyth-1-3837836, accessed 6 January 2019.

[263] Law, A., McLean, F. and Willby, N.H. (2016). Habitat engineering by beaver benefits aquatic biodiversity and ecosystem processes in agricultural streams. *Freshwater Biology*, 61(4), pp. 486–499. https://doi.org/10.1111/fwb.12721.

[264] WWF. (n.d.). *Brown Bear.* Worldwide Fund for Nature. http://www.worldwildlife.org/species/brown-bear, accessed 12 November 2016.

A study on the Kenai Peninsula, Alaska, found that brown bears (*Ursus arctos*) played a crucial role in recycling marine-derived nitrogen—up to 40 kilogrammes of nitrogen per year per bear—into forest ecosystems through predation on salmon and subsequent defaecation.[265] The study authors estimated that between 15.5 and 17.8% of the total nitrogen in spruce foliage within 500 metres of the stream was derived from salmon, 83–84% of this distributed by the bears. This is just one indication of the multiple roles that bears play in maintaining ecosystem productivity and functioning.

'Flagship' conservation species such as tigers, elephants and pandas, and large 'iconic' migratory fish species such as mahseer, sturgeon and Atlantic salmon, can also mobilise support for protection and restoration of the networks of interconnected habitats they depend upon to complete their life cycles,[266,267] with associated uplift in other species and linked ecosystem services beneficial to human communities.[268]

However, whilst pro-nature conservation initiatives have done much to safeguard vulnerable species by withholding important habitat from conversion to farmed, industrial and/or urban uses, it does not automatically follow that their outcomes are universally advantageous for all in society. Where narrowly framed, akin to other myopic forms of dedicated agricultural and industrial landscape uses, conservation outcomes can mask a range of disbenefits for overlooked ecosystem services. This is evidenced today by contemporary land claims across virtually the whole world, and in particular in former colonies including Australia, southern and east Africa, the USA and Canada where the historic rights of nomadic, endemic and other people were formerly largely suppressed under previous private resource ownership and neoliberal economic approaches.

---

[265] Hilderbrand, G.V., Hanley, T.A., Robbins, C.T. and Schwartz, C.C. (1999). Role of brown bears (*Ursus arctos*) in the flow of marine nitrogen into a terrestrial ecosystem. *Oecologia*, 121(4), pp. 546–550.

[266] Caro, T. (2010). *Conservation by Proxy: Indicator, Umbrella, Keystone, Flagship, and Other Surrogate Species*. Island Press.

[267] Everard, M. and Kataria, G. (2011). Recreational angling markets to advance the conservation of a reach of the Western Ramganga River. *Aquatic Conservation*, 21(1), pp. 101–108.

[268] Everard, M., Fletcher, M., Powell, A. and Dobson, M. (2011). The feasibility of developing multi-taxa indicators for freshwater wetland systems. *Freshwater Reviews*, 4(1), pp. 1–19.

Many nature reserves around the world claim to be the world's oldest, but a strong claimant to this title is Mihintale in Sri Lanka created in the third century BC. Like many older reserves, Mihintale was instigated as a hunting reserve for a local Sri Lankan royal family. The formation of the New Forest in southern England followed a similar pattern, William the Conqueror proclaiming it as a royal forest in about 1079 following the Norman Invasion. This 'new forest' was reserved for the use of royal hunts, mainly for deer, at the expense of more than 20 small hamlets and isolated farmsteads, and with any interference with the king's deer and its forage subject to heavy punishment. Various other older reserves, many of them now prestigious, also began life as areas annexed in a heavy-handed manner for the restricted use of privileged sectors of society, to the exclusion of other inhabitants or potential users, raising searching questions of 'environmental racism' (See Box 3.44). It still remains commonplace for the costs of conservation to be concentrated on the rural poor of less developed countries, with the benefits enjoyed by global elites.[269] Furthermore, conservation measures are generally dedicated to protect 'charismatic megafauna', yet many vital ecosystem processes remain not only poorly understood but dependent on organisms that are either not known or poorly characterised.

Box 3.44 Environmental Racism and the Establishment of National Parks

The Kruger National Park in South Africa supports a diversity of wildlife and associated ecosystem services, constituting a truly world-class National Park. However, its initial establishment in 1898 was achieved through mass clearances of tribal people who had lived on the land for uncounted generations. Many of the descendants of these displaced people still live in abject poverty in settlement camps adjacent to the park's perimeter.

This same pattern is replicated in the USA, where many Native Americans were expelled from land subsequently to be designated as National Parks.[270]

---

[269] Balmford, A. and Whitten, T. (2003). Who should pay for tropical conservation, and how could these costs be met? *Oryx*, 37(2), pp. 238–250.

[270] Merchant, C. (2007). *American Environmental History: An Introduction*. Columbia University Press, New York.

One such example is the establishment of Yellowstone National Park after it was signed into law in 1872, entailing the subsequent eviction, and in some cases killing, of hundreds of Native Americans, with the authorities of the time also reneging on previous treaty promises.[271]

Attempts continue today to move people from protected areas in Thailand, Botswana, Ethiopia, Tanzania, South Africa and India, entailing not only the movement of people but also the destruction of economies and historic associations with the land.[272]

From an ecosystems perspective, conservation policies and measures require continuous reinterpretation in terms of both their intended and their potential unintended outcomes. The same is also true for re-evaluation of outcomes from a socio-ecological point of view. A more recent development making progress with these measures has been a trend towards addressing nature reserves and nature conservation strategies not merely in terms of a limited focus on prescribed species and habitats, but also wider ecosystem processes and societal benefits expressed as ecosystem services (see Box 3.45). This broader, systemic approach not

Box 3.45 Evolving Approaches to Nature Conservation Taking Account of Wider Contributions to Ecosystem Services

In 2018, Natural England, the nature conservation agency for England, published the report *Accounting for National Nature Reserves*.[273] This report took a Natural Capital Accounting approach to some of England's most important habitats, species and geology under National Nature Reserve (NNR) designations, cumulatively covering approximately 0.7% of England's land surface. The report used an extended balance sheet displaying the state of these natural assets, the ecosystem services that they provide, and their associated benefits and economic value. The report was produced to support improved and transparent decision-making. It concluded that *"Leaving the environment in a better state for future generations will require meaningfully linking financial decisions with environmental assets and benefits"* in the management of environmental assets to deliver public benefit over the long-term.

---

[271] Spence, M.D. (1999). *Dispossessing the Wilderness: Indian Removal and the Making of National Parks*. Oxford University Press, Oxford.

[272] Brockington, D., Duffy, R. and Igoe, J. (2009). *Nature Unbound: Conservation, Capitalism and the Future of Protected Areas*. Routledge, London.

[273] Natural England. (2018). *Accounting for National Nature Reserves: A Natural Capital Account of the National Nature Reserves Managed by Natural England*. Natural England, Peterborough.

The global, intergovernmental Ramsar Convention on Wetlands has for many years promoted ecosystem service assessment as a basis for wise use and systemic management of all wetlands, including Ramsar Sites designated under the Convention. In October 2018, the Ramsar Commission adopted the RAWES (Rapid Assessment of Wetland Ecosystem Services) approach under Ramsar Resolution XII.17[274] as a rapid and cost-effective method for the systematic assessment of ecosystem services provided by wetlands, to be applied at wetland sites globally.

In 2015, the Indian Institute of Forest Management published an *Economic Valuation of Tiger Reserves in India*,[275] which took an ecosystem services approach. The report noted that, although the underlying objective of establishing India's Tiger Reserves was to ensure continuing evolutionary processes for tigers in the wild, the reserves also provide a range of associated economic, social, cultural and spiritual benefits (Fig. 3.17). These diverse ecosystem service benefits were estimated by both quantitative and qualitative means in order to improve the visibility of the economic and social benefits provided by Tiger Reserves, and as an evidence base for more targeted investment and management. Up to 25 services were addressed and valued for six of India's Tiger Reserves. The study found that a large proportion of benefits were intangible, and so were not reflected in market transactions. Yet, maintaining these flows of ecosystem service benefits to society is essential for the sustainability of the regions in which the Tiger Reserves are located. Evidence in the report also supported the need for enhancing or restoring ecological connectivity between Tiger Reserves.

2010 saw publication in the UK of the 'Lawton Review', *Making Space for Nature*,[276] addressing how England's wildlife and ecological network could be improved to help nature thrive in the face of climate change and other pressures. The report called for a *"bigger, better, more joined up"* approach, implicitly including increasing the suitability and permeability of non-designated landscapes for the movement of wildlife and flows of ecosystem processes and services, constituting a fundamental support system for human wellbeing. This requires the planning of ecological networks, including areas for restoration, better to optimise the multiple ecosystem services stemming from land use including, as examples, natural solutions to flood management.

[274] Ramsar Convention. (2018). *Resolution XIII.17: Rapidly Assessing Wetland Ecosystem Services.* 13th Meeting of the Conference of the Contracting Parties to the Ramsar Convention on Wetlands. https://www.ramsar.org/about/cop13-resolutions, accessed 4 March 2019.

[275] Verma, M., Negandhi, D., Khanna, C., Edgaonkar, A., David, A., Kadekodi, G., Costanza, R. and Singh, R. (2015). *Economic Valuation of Tiger Reserves in India: A Value+ Approach.* Indian Institute of Forest Management, Bhopal, India. https://conservewildcats.org/wp-content/uploads/sites/5/WildCats/papers/NTCA_Report2015.pdf, accessed 4 March 2019.

[276] Lawton, J.H., Brotherton, P.N.M., Brown, V.K., Elphick, C., Fitter, A.H., Forshaw, J., Haddow, R.W., Hilborner, S., Leafe, R.N., Mace, G.M., Southgate, M.P., Sutherland, W.J., Tew, T.E., Varley, J. and Wynne, G.R. (2010). *Making Space for Nature: A Review of England's Wildlife Sites and Ecological Networks.* Report to the Department for Environment, Food and Rural Affairs (Defra), London.

Fig. 3.17 A mature male tiger in Ranthambhore Tiger Reserve, Rajasthan, one of India's Tiger Reserves for which an ecosystem services assessment has been conducted to demonstrate the breadth of benefits they provide to society. (Image © Mark Everard)

only provides a greater societal rationale for continued support for conservation initiatives, but also enlightens how conservation outcomes, both with and beyond reserve boundaries, can best be achieved.

3.8   'Rewilding'

As an antidote to the widespread displacement of wildlife and ecosystem processes from increasingly developed landscapes, there is growing interest in 'rewilding'. Essentially, rewilding entails allowing natural succession to occur by radically reducing human interventions. It may also include reintroductions of formerly extirpated species, particularly top predators such as lynx, wolves and bears. This interest in rewilding spans a range of interests from the altruistic to the restoration of ecosystem functioning.

'Rewilding' is therefore both an attractive yet also an elusive concept. 'Wild' is generally defined by absence of human intervention, but what does it actually mean in a densely populated world wherein even our most extensive nature reserves, particularly across Africa and North America, are hardly free from human interventions? If we struggle with knowing what is genuinely 'wild', defining rewilding surely presents even greater conceptual challenges.

There are many global examples of inadvertent rewilding. For example, in the demilitarised 'no go' zone between North and South Korea, populations of rare birds such as red-crowned and white-naped cranes, mammals including the Amur goral, Asiatic black bear and musk deer, and other wildlife prosper in the absence of human presence and interventions. Also, a silver lining to the dark and radioactive cloud of the 1986 Chernobyl nuclear reactor disaster has been the resurgence of animals including lynx and European bison in the exclusion zone. Undoubtedly, much of nature would fare better if its principal pressures—human activities—were expunged.[277]

That we need to find more space for nature and natural processes in an ever more populated and densely developed world is certain. This is not only due to deep concerns about the ongoing viability of nature, but also for the continuing wellbeing of humanity. To quote IPBES in relation to its alarming assessment of sharply declining biodiversity, *"We are eroding the very foundations of our economies, livelihoods, food security, health and quality of life worldwide".*[278] In practical terms then, rewilding is necessarily a deliberate and active process of intervention in populated landscapes, intended to support the recovery of natural species and ecosystem processes in regenerating soils, water systems and their natural functions. The term 'rewilding' first occurred in print in 1990[279] with various subsequent redefinitions, including recognising the central role of *"cores, corridors, and carnivores".*[280]

---

[277] Wiseman, A. (2007). *The World Without Us*. Barnes and Noble, New York.

[278] Brondizio, E.S., Settele, J., Díaz, S. and Ngo, H.T. (2019). *Global Assessment on Biodiversity and Ecosystem Services of the Intergovernmental Science-Policy Platform on Biodiversity and Ecosystem Services (IPBES)*. IPBES. https://www.ipbes.net/global-assessment-biodiversity-ecosystem-services.

[279] Foote, J. (1990). Trying to take back the planet. *Newsweek*, 5 February 1990, p. 24.

[280] Soulé, M. and Noss, R. (1998). Rewilding and biodiversity: Complementary goals for continental conservation. *Wild Earth*, 8, pp. 19–28.

The Knepp Estate is a pioneering example of rewilding, or at least a managed version of rewilding, in a constrained lowland Britain context[281] (Box 3.46). Knepp's rewilding approach was innovated within the restrictions imposed by its densely farmed and populated surrounding landscape. The conservation ethos at what is now the Knepp Wildland Project differs radically from conventional nature conservation approaches, which tend to be predicated on preservation of habitats often in mid-succession to support specific species and other goals. However, efforts at Knepp Estate are process-led, driven by reestablishment of a functioning ecosystem granting nature as much freedom as possible within landscape constraints. The most extensive and longest-established British rewilding initiative is led by Reforesting Scotland[282] (Box 3.47). Other large land-owners in Scotland are keen to reintroduce wolves and lynx as 'lost' top predators,[283] though currently face opposition from planners and the wider population. Other significantly successful rewilding examples globally include an America project aiming to restore the prairie grasslands of the Great Plains (Box 3.48), as well as in Australia particularly where led by the NGO Rewilding Australia.[284]

Box 3.46 The Rewilding of the Knepp Estate in Sussex, England

The Knepp Estate spans 3,500 acres (1,400 hectares) approximately 8 km to the south of the town of Horsham in West Sussex, southern England. Ownership of the estate has been in the same family since 1802, during which time it has been subject to significant changes in farming regime responding to market, wartime and other pressures. Situated on 320 m of Wealden clay, the landscape itself makes for difficult farming. In the face of increasing intensification and chemical inputs in farming practices particularly during the 1980s and 1990s, the marginal nature of the land meant that arable and dairy farming at Knepp were becoming increasingly uncompetitive. At the same time, input intensity was contributing to ongoing precipitous declines in wildlife across all taxonomic groups. It is against this background that the owners decided in the mid-1990s to break away from conventional commodity farming, taking advice to migrate to a form of extensive ranching more closely akin to systems in Africa.

---

[281] https://knepp.co.uk/.

[282] http://www.reforestingscotland.org/what-we-do/the-reforesting-scotland-vision/.

[283] Hetherington, D., Cairns, P. and Geslin, L. (n.d.). *Could We Live with LYNX?* The Big Picture, Scotland. https://www.scotlandbigpicture.com/Downloads/STBP-stories-lynx.pdf.

[284] https://rewildingaustralia.org.au/.

Since 2001, this once intensively farmed landscape has instead been devoted to a pioneering rewilding project emulating processes in fully natural systems.[285] Using domestic grazing animals such as longhorn cattle, Tamworth pigs and Exmoor ponies as surrogates for their once pervasive wild forebears, in addition to native deer, natural landscape processes have been allowed to recover. Grazing pressures of migrating herds of animals have been key drivers of habitat creation and diversification, including natural restoration of dynamic watercourses.

Pigs rootle and turn over soil, cattle and deer graze and browse on trees and shrubs outdoors throughout the whole calendar year. These semi-natural grazing regimes with different animals free to move across the estate landscape stimulate widely differing vegetated zones, ranging from those that are densely wooded to savannah-like glades. This in turn, supported by cessation of pesticide inputs, saw rapid recovery of insect and bird populations. Some predicted but many more unanticipated ecological outcomes have resulted, including increases in scarce species such as turtle doves, nightingales, peregrine falcons and purple emperor butterflies, as well as booming populations of more common species.

Clearly, re-establishment of large predatory species lost from the British landscape would not be permissible in this densely-populated region of south-east England, so managed culling of herbivore stock is necessary to avoid over-grazing. This otherwise largely hands-off approach to ecological restoration is low-cost, and may be suitable for many more marginal or failing areas of farmland, building upon and connecting networks of established nature reserves across broad landscapes.

As a privately owned estate, the economics of nature-based framing have to stack up. Clearly, less food is produced compared to intensive farms using input-intensive methods. However, the culling of free-range grazing animals supports a market in organic, pasture-fed free-range meat, the UK government's Farm Business Survey[286] finding significantly higher financial returns at Knepp for lowland grazing livestock relative to the national average. Farm support payments, including the Basic Payment and supplementary Environmental Stewardship subsidy payments, make significant contributions. Farm labourer and building needs have also substantially reduced, freeing up housing and office/business space for rent. In addition, Knepp Estate has opened up wildlife safari, touring and camping/glamping operations, enabling people to experience the rich wildlife directly. The overall result, as revealed in the Savill's Rural Estate Benchmarking Survey,[287] is that like-for-like agricultural income at Knepp Home Farm outperforms the national average, excluding additional rental, tourism and other non-farm income.

---

[285] Tree, I. (2018). *Wilding: The Return of Nature to a British Farm.* Picador.

[286] https://www.gov.uk/government/collections/farm-business-survey.

[287] https://www.savills.co.uk.

Box 3.47 Reforesting Scotland

Reforesting Scotland, initiated in 1991, recognises that Scotland was once a well-forested country, its culture tied closely to trees and woodlands. However, multiple development and climate-related pressures have today decimated tree cover. Reforesting Scotland aims to reverse this situation, restoring a well-forested, productive landscape and its associated culture and economy.

Unlike Knepp Estate, Reforesting Scotland does not own land but is a membership organisation concerned with ecological and social regeneration, united by the common aim of significantly increasing the forested areas of Scotland.

Box 3.48 Rebuilding an American Prairie Reserve

The American Great Plains are naturally occurring grassland, significantly shaped by the grazing activities of large herbivores. However, they were disastrously degraded by deep and extensive tillage driven by economic stimuli during the Great Depression, inadvertently promoting the massive erosion behind the Dust Bowl disaster that drove 3.5 million from Plains states between the 1930s and 1940s.

The American Prairie Foundation, a charity based in the US state of Montana, aims to rebuild an American Prairie Reserve.[288] The vision for this Reserve is that it will constitute one of the largest wildlife reserves in the continental United States. This is to be achieved through a combination of new land acquisition and public land integration. Part of the plan entails reintroduction of bison on private land in the Missouri Breaks, emulating former habitat-forming grazing regimes.

Clearly, restoration of populations of large predators may be problematic in areas of dense human settlement and agricultural uses. However, it can be effective in more remote upland and nature reserve areas, where restoration of ecosystem functioning is a priority and can deliver a diversity of benefits into adjacent areas. As previously observed, surprising results have ensued from reintroduction of the grey wolf into several areas in the northern Rocky Mountains of the United States since 1995, including in Yellowstone National Park, driving fundamental changes in

---

[288] https://www.americanprairie.org/.

the character and functioning of these ecosystems. This represents a form of beneficial rewilding in a region that had formerly already been considered a wilderness. Also, reintroduction of beavers in Devon and other parts of the UK and US are restoring ecosystem processes, giving rise to improved biodiversity, fisheries, hydrology and water quality through their role as 'ecosystem engineers' as a form of rewilding.

Rewilding of marginal farmed land can offer advantages simultaneously to wildlife and ecosystem processes, but also to farming and land management enterprises though the latter are not without risks. For example, whilst less vulnerable to fluctuations in commodity prices, successes at Knepp Estate are subject to uncertainties in both farming and nature conservation legislation. Potential conflicts with agricultural legislation arise from the fact that wildlife-rich scrub habitat supporting free-range grazing and browsing might fail to meet current legal conditions about maintenance of land in 'good condition' as defined for intensive farming systems. Concerns about nature conservation payments revolve around their current focus on retaining woodland, meadow, scrub or other habitats in static condition, rather than allowing dynamism in the landscape through process-based conservation.

These concerns highlight the importance of a supportive policy and financial environment to enable rewilding, in part or in full, and other forms of nature-based regeneration. Emerging political rhetoric around refocussing rewards for land management away from support for commodity production and instead towards natural capital conservation and flows of ecosystem services has yet to translate substantially into a transformed payment regime stimulating a shift in the mainstream of land management. Rewilding also challenges the established goal-driven frameworks underpinning much contemporary conservation work, which is generally predicated on specific scarce species rather than promoting underlying dynamic ecosystem processes. A process-based approach may have less predictable outcomes, yet has a higher likelihood of multi-beneficial results not only across taxonomic groups but also for regeneration of societally valuable ecosystem services. These societal benefits may potentially include contributions, in suitable landscapes, towards the goal of feeding the growing global human population profitably yet with less carbon intensity, habitat displacement and other pressures on the natural world.

Whilst restoration of the full complement of native fauna, including top predators, is emotionally appealing and would undoubtedly yield significant and long-lasting ecosystem service benefits, it is far from practicable everywhere. Large predators 'over-spilling' from rewilded areas to roam urban settlements and traditional farmland would be unwelcome. Furthermore, for all its negative side effects, we also need intensive farming as part of a mix of land uses to feed the demands of an increasingly populated world. Conversely, where farming is of marginal profitability, as was formerly the case at Knepp Estate and is commonly so elsewhere in whole farming estates as well as areas of larger farm units, rewilding is an approach yielding many potential benefits to the farm, wildlife, the wider population and the diverse needs of current and future society.

A key issue underpinning the thinking behind, and practice of, rewilding is that humanity and nature are interdependent. The massive global declines in biodiversity reported by IPBES are of far more than altruistic concern, also relating to the viability, economic progress and quality of life of current and future human generations. So rewilding is, in essence, as much about reintegrating ecosystems and their beneficial processes back not only into our nature reserves and farmland, but also our built and otherwise developed environments, to recognise and realise that our health and wellbeing are intimately linked with the vitality of the Earth's ecosystems at all scales.

3.9   Farming for Human Wellbeing

A regenerative approach applies not just in remote or conserved landscapes, but to all populated landscapes and their supportive processes. The actual as well as potential contributions of these processes need to be recognised, used wisely and protected, enhanced or, in cases such as constructed wetlands, emulated to continue to support continuing human interests on a sustainable basis. Farming landscapes are frequently in decline as intensive tillage and other practices maximise a limited subset of services generally related to commodity production, commonly to the detriment of ecosystem integrity and production of a breadth of beneficial services. Many examples of unsustainable outcomes from intensification of agriculture have been identified throughout this book, highlighting

Fig. 3.18   Mainstream intensive farming does not automatically work in synergy with all ecosystem processes. (Image © Dr Mark Everard)

the need for wide-scale change in our stewardship of landscapes. Farmed land offers significant opportunities for sustainable change, given that some 16,000 million hectares are farmed globally though this area has been declining since the last century (Fig. 3.18).[289]

Unsurprisingly, substantial global attention is being paid to sustainable agriculture. Much of the scientific discourse relates to the trade-offs and synergies in optimising ecosystem service benefits[290] to ensure their continued provision whilst addressing the challenge of producing 50% more food for nine billion people by 2050.[291] There is growing recognition that understanding the whole socio-ecological system at local to landscape scales is required to make a balanced assessment of ecosystem services

[289] Ausubel, J.L., Wernick, I.K. and Waggoner, P.E. (2013). Peak farmland and the prospect for land sparing. *Population and Development Review*, 38(s1), pp. 221–242.

[290] Power, A.G. (2010). Ecosystem services and agriculture: Tradeoffs and synergies. *Philosophical Transactions of the Royal Society B—Biological Sciences*, 365(1554). https://doi.org/10.1098/rstb.2010.0143.

[291] FAO. (n.d.). *Ecosystem Services Sustain Agricultural Productivity and Resilience*. UN Food and Agriculture Organization, Rome. http://www.un.org/esa/sustdev/csd/csd16/documents/fao_factsheet/ecosystem.pdf, accessed 6 November 2016.

provided by agriculture.[292] Nevertheless, despite high-level commitments to achieve desired outcomes at landscape scale, practical implementation requires an integrated set of 'societal levers' to influence the decisions of land managers around the achievement of clear, multi-beneficial goals.[293] These 'levers' comprise a wide variety of formal and informal elements of the policy environment, ranging from market forces, statutory legislation, common/civil law, market-based instruments including subsidies, taxes and other means, and protocols influencing land use; these currently fall well short of matching high-level aims to achieve sustainability.[294] Solutions need to be systemically informed, but also practical and workable for farming interests.

There are some exemplars of best practice in achieving 'regenerative landscapes' that can be helpful in influencing the wider policy environment, and promoting uptake and evolution of sustainable farming methods that address ecosystem integrity and the protection of multiple ecosystem services. Progress at the UK-based Loddington Farm (Box 3.49) and Hope Farm (Box 3.50), and in the US at the Kellogg Biological Station (Box 3.51) is illuminating. All three case studies highlight how targeting of multiple beneficial ecosystem service outcomes, rather than maximising commodity production in isolation, is not only feasible but also profitable. It has also been achieved using 'off the shelf' technologies applied with forethought. Successes at Loddington Farm, Hope Farm and the Kellogg Biological Research Station highlight tractable and profitable progress with commodity production in conjunction with multiple beneficial outputs, with ecosystem integrity, biodiversity and resilience as central features (Fig. 3.19).

---

[292] Robertson, G.P., Gross, K.L., Hamilton, S.K., Landis, D.A., Schmidt, T.M., Snapp, S.S. and Swinton, S.M. (2014). Farming for ecosystem services: An ecological approach to production agriculture. *BioScience*, 64, pp. 404–415. https://doi.org/10.1093/biosci/biu037.

[293] Everard, M. (2011). *Common Ground: The Sharing of Land and Landscapes for Sustainability*. Zed Books, London, 214pp.

[294] Everard, M., Dick, J., Kendall, H., Smith, R.I., Slee, R.W., Couldrick, L., Scott, M. and MacDonald, C. (2014). Improving coherence of ecosystem service provision between scales. *Ecosystem Services*, 9, pp. 66–74.

Box 3.49 Progress with Ecosystem Service Production at Loddington Farm

Loddington Farm in Leicestershire, eastern England, was bequeathed to the Allerton Project (initially the Allerton Research and Educational Trust) with the aims of advancing public education and conducting research on different farming methods and their effects on the environment and wildlife.

The Game & Wildlife Conservation Trust, a UK-based NGO, has promoted advances in game and wildlife management in a 333 ha commercial farm context under the Allerton Project. The Trust's approach embraces modern technologies for arable cropping, sheep grazing, and the maintenance of woodland, a stream and several ponds. Traditional tillage techniques reverted to minimal tillage in 1997 and to disc cultivator drilling in 2001, changing from tyres to tracks on the tractor to reduce soil compaction. Research also expanded to address soil erosion, organic matter, soil flora and fauna, and nutrient management. Direct drilling has since been adopted as the primary cultivation tool, using a lighter tractor to reduce soil compaction with benefits for greater fuel efficiencies and reduced exhaust emissions. Water-friendly farming measures were launched in 2012.[295]

Breeding songbird numbers recovered following implementation of a game management system in 1992, rapidly doubling though with some small decline then occurring after 2001 when winter feeding and predator control were withdrawn.[296] Reduced ploughing has also enabled worm populations to increase, improving soil permeability and retention of organic matter and nutrients.[297]

The Loddington experience demonstrates how economically profitable farming that also operates in greater synergy with natural processes, seeking linked outcomes for wildlife, water, nutrients and other ecosystem services, is feasible with off-the-shelf solutions.

---

[295] Biggs, J., Stoate, C., Williams, P., Brown, C., Casey, A., Davies, S., Grijalvo Diego, I., Hawczak, A., Kizuka, T., McGoff, E. and Szczur, J. (2014). *Water Friendly Farming: Results and Practical Implications of the First 3 Years of the Programme*. Freshwater Habitats Trust/Game & Wildlife Conservation Trust, Oxford/Fordingbridge. http://www.gwct.org.uk/media/434327/Water-Friendly-Farming-Report-2014a.pdf, accessed 6 November 2016.

[296] GWCT. (n.d.). *Allerton Project: Songbird Research*. Game & Wildlife Conservation Trust. http://www.gwct.org.uk/allerton/songbird-research/, accessed 19 May 2017.

[297] GWCT. (n.d.). *Wildlife: October: Worms*. Game & Wildlife Conservation Trust. https://www.gwct.org.uk/wildlife/species-of-the-month/2010/october/, accessed 19 May 2017.

Fig. 3.19 Reintegrating unmanaged habitat areas such as untilled field margins, hedges, copses and wetlands within farmed landscapes beneficially promotes pollinating species and the predators of crop pests. (Image © Dr Mark Everard)

Box 3.50 Combining Nature Recovery with Profitable Farming at Hope Farm

In 2000, the Royal Society for the Protection of Birds (RSPB: a UK-based nature conservation NGO) purchased Hope Farm, an arable farm in Cambridgeshire, for research and as a showcase and means to influence nature-friendly farming.[298] Total farm area is 181 hectares of which 161 hectares are cropped, 6 hectares are pasture and 0.5 of a hectare is woodland. The land is farmed on contract by a neighbouring farmer, with on-farm work paid at realistic commercial rates. The aim is not only to institute nature-friendly methods but to do so whilst remaining profitable. Much of UK wildlife depends on farmland, and around 75% of UK land is covered by agriculture, but intensive farming methods have also been one of the major negative pressures on biodiversity.

---

[298] RSPB. (2017). *Hope Farm: Farming for a Sustainable Future—For People and Wildlife.* Royal Society for the Protection of Birds (RSPB), Sandy. https://www.rspb.org.uk/globalassets/downloads/about-us/hope-farm-update-2017.pdf, accessed 8 May 2019.

Hope Farm, like Loddington Farm, acts as a demonstration site illustrating for other farmers of how to combine nature conservation with profitability. Changes in stewardship since purchase include new techniques to improve food and shelter for birds and soil quality through measures such as retention of overwinter stubble, growing post-harvest cover crops, and the protection and improvement of hedgerows, wetland features and other valuable habitat. Monitoring of farm wildlife between 2000 and 2017 found a 213% increase in the numbers of butterflies and a 226% increase in numbers of breeding farmland birds, whilst simultaneously maintaining profitability for both the landowner and contract farmer.

Box 3.51 Progress with Ecosystem Service Production at the Kellogg Biological Research Station

The Kellogg Biological Research Station, associated with Michigan State University, has been a Long-Term Ecological Research site of the US National Science Foundation. 25 years of experimentation reveal that a range of ecosystem services—clean water, biocontrol, biodiversity benefits, climate stabilisation and long-term soil fertility—can be provided by intensive row-crop ecosystems that also produce high agricultural yields.[299]

Midwest farmers, especially on large farms, were willing to adopt practices delivering these services in exchange for payments scaled to management complexity and farmstead benefit. Surveyed citizens also indicated a willingness to pay farmers for the delivery of specific additional services. A new, profitable 'farming for ecosystem services' paradigm in US agriculture seems attainable with appropriate policy evolution.

There is a long history of recognition that the farming intensification approaches since the latter half of the Twentieth Century require radical reform if sustainability is to be achieved, for example as set out in various UK reports and statements of policy direction (Box 3.52).

---

[299] Robertson, G.P., Gross, K.L., Hamilton, S.K., Landis, D.A., Schmidt, T.M., Snapp, S.S. and Swinton, S.M. (2014). Farming for ecosystem services: An ecological approach to production agriculture. *BioScience*, 64, pp. 404–415. https://doi.org/10.1093/biosci/biu037.

Box 3.52 UK Examples of Shifting Policy Thinking in about Farming

Various reports to government and ensuing statements of policy direction in the UK concerning an increasingly multi-beneficial approach to farming and land use include:

- The 2002 'Curry Report', *Farming and Food: a sustainable future*,[300] set out a vision for the food and farming industry that included a profitable and sustainable farming and food sector capable of competing internationally, but for which stewardship of the environment including public payments to maintain public benefits was a key features.
- The 2010 'Lawton Review', *Making Space for Nature*,[301] addressed previously in the context of nature reserves, also recognised the need to make wider landscapes more habitable and permeable for nature and more effective in providing the wide diversity of ecosystem services essential for the wellbeing of society under a *"bigger, better, more joined up"* approach.
- The 2018, UK government policy statement, *Health and Harmony*,[302] reiterated the importance of food, farming and the environment after the UK leaves the EU. The policy statement noted that *"British farmers, growers and foresters play a vital role in protecting the countryside, while producing world-class food, plants and trees"*. It addressed the use of public money to enrich wildlife habitats, prevent flooding, improve the quality of air, soil and peat, and to plant trees as a basis for providing public goods which help manage and mitigate the effects of climate change.
- UK government also set out a broad range of aspirations under its 2018 *A Green Future: Our 25 Year Plan to Improve the Environment*,[303] setting out a broadly framed blueprint for the environment for the coming quarter-century aimed at using a natural capital approach to build a healthier environment as a basis for sustainable growth and improving wellbeing. Farmed land is very much at the fore in this 25-year plan, subject to both new rules but also a reformed subsidy system geared to realisation of public benefits in return for public money.

---

[300] Curry, D. (2002). *Farming and Food: A Sustainable Future: Report of the Policy Commission on the Future of Farming and Food, January 2002.* Cabinet Office, London. https://webarchive.nationalarchives.gov.uk/20100702224742/http://archive.cabinetoffice.gov.uk/farming/pdf/PC%20Report2.pdf.

[301] Lawton, J.H., Brotherton, P.N.M., Brown, V.K., Elphick, C., Fitter, A.H., Forshaw, J., Haddow, R.W., Hilborner, S., Leafe, R.N., Mace, G.M., Southgate, M.P., Sutherland, W.J., Tew, T.E., Varley, J. and Wynne, G.R. (2010). *Making Space for Nature: A Review of England's Wildlife Sites and Ecological Networks.* Report to the Department for Environment, Food and Rural Affairs (Defra), London.

[302] Defra. 2018. *Health and Harmony: The Future for Food, Farming and the Environment in a Green Brexit—Policy Statement.* Department for Environment, Food and Rural Affairs (Defra), London.

[303] HM Government. (2018). *A Green Future: Our 25 Year Plan to Improve the Environment.* HM Government, London.

In practice, intensely farmed landscapes across the wider world have been seeing an increasing range of nature-based solutions. Many in Europe are driven by the Common Agricultural Policy (CAP), though many more internationally are relearning traditional practices, to reduce dependence on potentially damaging, chemically and energy-intensive farming practices (Box 3.53). These are, in the main, marginal rather than mainstream activities, albeit that some are eligible for public subsidies in some countries in recognition of their contributions to sustainability. Others in less developed and agriculturally intensive nations are 'business as usual' traditional practices honed over millennia to coexist with landscapes and benefit from nature's services. All demonstrate beneficial integration of natural systems, or their reintegration into a formerly over-mechanised and industrialised farming systems.

Box 3.53 Examples of Ecosystem-based Farming Solutions

- Hedgerows were once a common feature of farmed land across Europe but, to make way for increasingly large farm machinery and production intensity, were widely removed particularly during the 1980s. Now, their value is beginning to be appreciated, particularly where established along the contours of fields retaining water and soil, promoting percolation into groundwater and storing flood run-off, as habitat for pest predators and pollinators, and serving as wind breaks, amongst a range of beneficial services.[304]
- Riparian buffer zones, also known as stream buffers or buffer strips, are margins of watercourses protected from tillage or grazing where vegetation can regenerate (Fig. 3.20). These buffer zones stabilise river banks, prevent eroded soil entering streams, provide habitat for fish and particularly increasing spawning zones and nursery areas for fry, and support pollinators and the predators of crop pests.[305] Further direct farming benefits include helping prevent livestock straying, and also reducing hoof rot resulting from wallowing too frequently in damp river margins.
- Beetle banks comprise low earth banks built across the middle or alongside arable fields, innovated in England in the early 1980s with the aim of providing suitable overwintering habitat for predatory insects and spiders that move into the neighbouring crops reducing pest species sig-

---

[304] Pagella, T.F. and Sinclair, F.L. (2014). Development and use of a typology of mapping tools to assess their fitness for supporting management of ecosystem service provision. *Landscape Ecology*, 29(3), pp. 383–399.

[305] Everard, M. (2015). *River Habitats for Coarse Fish: How Fish Use Rivers and How We Can Help Them!* Old Pond Publishing, Sheffield.

nificantly. Beetle banks also play host to a variety of wildlife from nesting birds and bumblebees to hares, harvest mice, and wild flowers providing nectar for pollinators.

- Review of many case studies reveals that low fruit or seed set by crop species, and the resulting reduction in crop yields, are attributable to impoverishment of pollinator diversity.[306] Increasing evidence indicates that conserving wild pollinators in habitats adjacent to agriculture improves both the level and the stability of pollination services, leading to increased yields and income (Fig. 3.21).[307]

Fig. 3.20 Buffer zones preventing livestock access to rivers stop animals straying and reduce foot rot and other diseases related to animals standing in water, also allowing the regeneration of riparian habitat supporting a diversity of ecosystem processes and benefits. The fence excluding riparian grazing and trampling here on the Bristol Avon in Wiltshire is now almost obscured by dense naturally regenerated vegetation and multi-beneficial river bank habitat. (Image © Dr Mark Everard)

[306] Richards, A.J. 2001. Does low biodiversity resulting from modern agricultural practice affect crop pollination and yield? *Annals of Botany*, 88, pp. 165–172.

[307] Klein, A.M., Steffan-Dewenter, I. and Tscharntke, T. 2003. Fruit set of highland coffee increases with the diversity of pollinating bees. *Proceedings of the Royal Society B—Biological Sciences*, 270(1518), pp. 955–961.

Fig. 3.21 'Rough edges' to field, along hedge line and on headlands, can provide habitat for a diversity of species including serving as 'beetle banks' supporting pollinators and the predators of crop pests. (Image © Dr Mark Everard)

Utilisation of land to grow food and other commodities is obviously essential to support the needs of high, growing and intensely clustered human populations. However, the pattern of wholesale landscape change seen in mainstream intensive farming in Europe and the USA, eradicating natural features to make way for input-intensive monocultures, is far from the only means to achieve this, and may in fact not be the most beneficial. As one example, paludiculture practices make use of wetland soils for the production of biomass whilst retaining or reintroducing moisture simultaneously to maintain peat and stored carbon.[308] A similar approach is taken by permaculture, the term

[308] Wichtmann, W., Haberl, A. and Tanneberger, F. (2010). *Production of Biomass in Wet Peatlands (Paludiculture)*. The EU-AID project 'Wetland energy' in Belarus—Solutions for the substitution of fossil fuels (peat briquettes) by biomass from wet peatlands. Michael Succow Foundation, Greifswald, Germany.

concatenating 'permanent agriculture',[309] comprising a set of design principles making use of features of natural ecosystems and working with, rather than against, natural processes. Permaculture is not defined by uniform methods, as often featuring in contemporary intensive practices, but is most commonly articulated as following twelve principles (Box 3.54).

Box 3.54 Twelve Principles of Permaculture, as Defined in 2002 by David Holmgren[310]

1. *Observe and interact*: By taking time to engage with nature we can design solutions that suit our particular situation.
2. *Catch and store energy*: By developing systems that collect resources at peak abundance, we can use them in times of need.
3. *Obtain a yield*: Ensure that you are getting truly useful rewards as part of the work that you are doing.
4. *Apply self-regulation and accept feedback*: Discourage inappropriate activity to ensure that systems can continue to function well.
5. *Use and value renewable resources and services*: Make the best use of nature's abundance to reduce our consumptive behaviour and dependence on non-renewable resources.
6. *Produce no waste*: By valuing and making use of all the resources that are available to us, nothing goes to waste.
7. *Design from patterns to details*: By stepping back, we can observe patterns in nature and society. These can form the backbone of our designs, with the details filled in as we go.
8. *Integrate rather than segregate*: By putting the right things in the right place, relationships develop between those things and they work together to support each other.

---

[309] Paull, J. (2011). The making of an agricultural classic: Farmers of forty centuries or permanent agriculture in China, Korea and Japan, 1911–2011. *Agricultural Sciences*, 2(3), pp. 175–180.

[310] Holmgren, D. (2002). *Permaculture: Principles and Pathways Beyond Sustainability*. Permanent Publications, East Meon.

9. *Use small and slow solutions*: Small and slow systems are easier to maintain than big ones, making better use of local resources and producing more sustainable outcomes.
10. *Use and value diversity*: Diversity reduces vulnerability to a variety of threats and takes advantage of the unique nature of the environment in which it resides.
11. *Use edges and value the marginal*: The interface between things is where the most interesting events take place. These are often the most valuable, diverse and productive elements in the system.
12. *Creatively use and respond to change*: We can have a positive impact on inevitable change by carefully observing, and then intervening at the right time.

One widespread farming system that works with natural landscape character is the terraced rice paddies that are widespread across Asia, persisting in some places for over six millennia and substantially outliving civilisations that have relied upon them. Terraced agriculture is pervasive in sloping landscapes throughout Asia, where the overwhelming majority of rice is still produced within walking distance of where it is eaten.[311] Wheat production tends to replace rice in terraced farming systems at higher, cooler altitudes. Though 'low tech' in western industrial terms, terraced cultivation systems founded on centuries-old traditional knowledge work with rather than fundamentally change natural landscape character, and are remarkably efficient at conserving not only water but also soil and nutrients (Fig. 3.22).

Polyculture is also widely practiced, for example by the introduction of small fish when terraced rice paddy ponds are flooded. The fish are then harvested when the paddy is drained, providing an important local source of protein. There are many other examples of effective polyculture systems working in sympathy with natural processes, such as the multi-level

---

[311] Codrington S. (2005). *Planet Geography*. Solid Star Press, North Ryde.

Fig. 3.22 Terraced farming is an ages-old practice, efficiently retaining water, nutrients and soil in sloping terrain across Asia. (Image © Dr Mark Everard)

vegetative structure found in India's coffee groves (Box 3.55) (Fig. 3.23) or taller trees providing shade in tea plantations.

Box 3.55 Polyculture in the Coffee Groves of Coorg, Karnataka

Coffee is a key crop produced in the higher elevations of Coorg region in the Indian state of Karnataka. Here, coffee estates use a multi-layered polyculture system, yielding multiple crops.

Coffee bushes require shade, which is provided by a diversity of trees that also yield useful products. These include, for example, silver oak (for timber), jackfruit (for edible fruit), kapok (for fibre) and cashew. The trees serve as supports for vines producing vanilla and pepper. An understory of salad and other vegetable crops is common, mainly providing for domestic needs rather than grown as commodity crops.

The microclimate within coffee groves is substantially cooler than the surrounding open farmlands, and dense mists tend to form in the morning as evaporation condenses within tight local water cycles. This microclimate under the tree canopy is commonly well populated by birds, butterflies and other insects, with leaf fall retaining nutrients in the soil.

The use of shade by retention or planting of trees in pasture is common in regions such as East Africa where direct and strong sunlight can suppress the growth of grass. However, the linked benefits of silvopasture, also known as agroforestry, are being increasingly recognised elsewhere, including across Europe and in the US (Box 3.56).

Box 3.56 The Silvopasture Approach

Silvopasture comprises the intensive management and growth of perennial grasses, or grass-legume mixes, in a forest stand that provides a more naturalistic form of livestock grazing emulating natural interactions between herbivores and mixed vegetation.[312] This integrated approach, including rotational grazing, can simultaneously optimise grazing animal and timber production. Tree growth is enhanced by reductions in competition for moisture, nutrients and sunlight and by natural fertilisation, whilst meat and milk production are enhanced by windbreak shelter and tree fodder. In addition to on-farm benefits, including reduced soil compaction, erosion and local recirculation of water, wider off-farm public benefits arise through addressing downstream problems of flood run-off and siltation, improved carbon storage and protection of biodiversity.

Though labour-intensive, silvopasture systems reduce economic risks through production of multiple products. They also require far less energy and agrochemical inputs than input-intensive monoculture approaches, with the additional production of off-farm, public benefits that may attract subsidies or other forms of payment.

No single solution is a panacea to the conflict inherent in the term 'sustainable intensification'. The term 'sustainable intensification' is something of a 'holy grail' vision, widely varying and disputed in definition, relating to resolving the conflicting pressures of feeding a growing population whilst protecting or restoring ecosystems. However, we are witnessing an expanding toolkit, many approaches reflecting traditional stewardship practices. A diversity of methods takes an ecosystem-based approach to deliver not simply a narrowly framed set of food or other commodity outputs but that also reflecting the multifunctionality of

---

[312] Angima, S.D. (2009). *Silvopasture: An Agroforestry Practice*. Oregon State University, EM 8989-E. https://catalog.extension.oregonstate.edu/sites/catalog/files/project/pdf/em8989.pdf, accessed 11 February 2019.

Fig. 3.23 Mist forming at dawn under the tree canopy in a multi-level, polyculture coffee grove in Coorg, in the uplands of the Indian state of Karnataka. (Image © Dr Mark Everard)

land. Rather than fighting natural processes, these novel or traditional methods seek to use natural processes and services such as pollination and pest control, soil protection and regeneration, nutrient cycling, wind and storm buffering, water and carbon retention, and permeability for infiltration of water. This has the hallmarks identified as 'systemic solutions, "…*low-input technologies using natural processes to optimise benefits across the spectrum of ecosystem services and their beneficiaries*", if all benefits in addition to commodity outputs are recognised by those taking decisions and by the policy and fiscal environment. These and other novel, ecosystem-based farming approaches can therefore potentially make strategic contributions to 'sustainable intensification' through choices and implementation nuanced to local geography, land suitability and societal needs.

A global study published in 2018[313] recognised that sustainable intensification of agricultural systems offers synergistic opportunities for the co-production of agricultural and natural capital outcomes. The study outlined progress towards sustainable intensification measured, both by farm and by hectare, by implementation of seven sub-type methods: integrated pest management; conservation agriculture; integrated crop and biodiversity; pasture and forage; trees; irrigation management; and small or patch systems. 29% of all farms worldwide were found to be practising at least one form of sustainable intensification on around 9% of global agricultural land area. This led the report's authors to conclude that sustainable intensification may be approaching a 'tipping point' where it could be transformative. What is certain is that a radical change in global norms is required given the massive scale (16,000 million hectares) of land area farmed globally, potentially driving the kinds of transformations into whole regenerative socio-ecological landscapes witnessed in China's Loess Plateau, the Ethiopian Highlands, the Everglades and Ireland's Anne Valley.

---

[313] Pretty, J., Benton, T.G., Bharucha, Z.B., Dicks, L.V., Flora, C.B., Godfray, H.C.J., Goulson, D., Hartley, S., Lampkin, N., Morris, C., Pierzynski, G., Prasad, P.V.V., Reganold, J., Rockström, J., Smith, P., Thorne, P. and Wratten, S. (2018). Global assessment of agricultural system redesign for sustainable intensification. *Nature Sustainability*, 1, pp. 441–446.

3.10   In the City

In 2018, 55% of the global population lived in cities with an anticipated rise to 68% by 2050 as urban populations grow by 2.5 billion, 90% of this increase taking place in Asia and Africa.[314] Reintegrating ecosystems and their diverse supportive processes into built environments then becomes increasingly essential if we are not to starve ourselves of their many benefits, or to suck the life from outlying hinterlands in our hunger and thirst for resources of all kinds (Fig. 3.24).

There is growing interest and application of 'green infrastructure', a term spanning diverse techniques integrating or emulating natural processes, reconciling environmental needs with economic growth in urban settings.[315] Some common examples are outlined in Box 3.57, with a more specific focus on New York City's street trees in Box 3.58. Green areas, vegetation and trees in urban areas have been linked to direct health

Fig. 3.24   Humanity is increasingly urban, Yokohama merging with Tokyo (in the right background) creating one of the densest urban centres on Earth. (Image © Dr Mark Everard)

---

[314] United Nations. (2018). *Revision of World Urbanization Prospects*. United Nations Department of Economic and Social Affairs. https://www.un.org/development/desa/publications/2018-revision-of-world-urbanization-prospects.html, accessed 10 January 2019.

[315] Horwood, K. (2011). Green infrastructure: Reconciling urban green space and regional economic development: Lessons from experience in England's north-west region. *Local Environment*, 16, 963–975.

benefits, including both mental health[316] as well as biophysical health (Fig. 3.25), as for example in a study from New York correlating the presence of street trees with a significantly lower prevalence of early childhood asthma.[317] Green area accessibility has also been linked to reduced mortality[318] and improved perception of general health.[319] These ecosystem-based technologies can represent multi-beneficial 'systemic solutions' for cities, contributing to economic development, wealth generation, habitat for nature and human wellbeing, and with far lower cost and emissions than single-purpose, mechanical solutions (Fig. 3.26).[320]

Box 3.57 Examples of Urban Green Infrastructure

Widely implemented examples of urban green infrastructure solutions include green roofs, WSUD (water-sensitive urban design), green walls, rain gardens, community forests, urban river restoration and rainwater harvesting, all harnessing ecosystem processes to contribute to reduced management costs and a higher quality urban environment.[321,322]

Prominent amongst there urban solutions are sustainable drainage systems (SuDS). SuDS themselves encompass a diversity of techniques[323] ranging from urban filter drains and pervious pipes, which produce minimal or no ecosystem service benefits beyond local flood regulation, through to constructed wetlands optimally designed to produce a diversity of regulatory and cultural benefits in addition to providing limited habitat for wildlife (supporting services) (Fig. 3.27).

In more rural settings, Rural Sustainable Drainage Systems (RSuDS) are implemented to reduce transport of pollutants to watercourses and offset

---

[316] Nilsson, K., Sangster, M., Gallis, C., Hartig, T., de Vries, S., Seeland, K. and Schipperijn, J. (2104). *Forests, Trees and Human Health*, Springer.

[317] Lovasi, G.S., Quinn, J.W., Neckerman, K.M., Perzanowski, M.S. and Rundle, A. 2008. Children living in areas with more street trees have lower prevalence of asthma. *Journal of Epidemiology and Community Health*, 62: 647–649.

[318] Mitchell, R. and Popham, F. (2008). Effect of exposure to natural environment on health inequalities: An observational population study. *Lancet*, 372(9650), pp. 1655–1660.

[319] Maas, J., Verheij, R.A. and Groenewegen, P.P., et al. (2006). Green space, urbanity, and health: How strong is the relation? *Journal of Epidemiology and Community Health*, 60(7), p. 587

[320] Habitat UN. (2012). *Urban Patterns for a Green Economy: Working with Nature*. UNON, Publishing Services Section, Nairobi, Kenya.

[321] Grant, G. (2012). *Ecosystem Services Come to Town: Greening Cities by Working with Nature*. Wiley-Blackwell, Chichester, UK.

[322] Everard, M. and Moggridge, H.L. (2012). Rediscovering the value of urban rivers. *Urban Ecosystems*, 15(2), pp. 293–314.

[323] Woods-Ballard, B., Kellagher, R., Martin. P., Jefferies, C., Bray, R. and Shaffer, P. (2015). *The SUDS Manual*. CIRIA Report C753. Construction Industry Research and Information Association (CIRIA), London.

other damage from farmed landscapes.[324] RSuDS include, for example, vegetated riparian strips, swales, interception ponds, buffer zoning, and rehabilitation or protection of habitat such as wetlands that are critical for
provision of various ecosystem services.

Perhaps the most advanced and pervasive example of green infrastructure deployment occurs in the densely populated small island city-state of
Singapore. Here, multiple techniques reintegrate various formerly displaced
ecosystem services, contributing to climate resilience, emission reductions,
balanced water flows and thermal comfort,[325] including rooftop farming to
address food security and carbon footprint concerns.[326]

Fig. 3.25 Mature street trees in even the densest city centres, as here in the
Harajuku district of Tokyo, Japan, provide multiple environmental and cultural
benefits as part of urban green infrastructure. (Image © Dr Mark Everard)

[324] Avery, L.M. (2012). *Rural Sustainable Drainage Systems (RSuDS)*. Environment Agency, Bristol.

[325] Demuzere, M., Orru, K., Heidrich, O., Olazabal, E., Geneletti, D., Orrugh, H., Bhave, A.G., Mittal, N., Feliu, E. and Faehnle, M. (2014). Mitigating and adapting to climate change: Multi-functional and multi-scale assessment of green urban infrastructure. *Journal of Environmental Management*, 146, 107–115.

[326] Astee, L.Y. and Kishnani, N.T. (2017). Building integrated agriculture: Utilising rooftops for sustainable food crop cultivation in Singapore. *Journal of Green Building*, 5(2), 105–113.

Box 3.58 Valuing New York Street Trees

New York City (NYC) is one of a number of global cities at the forefront of recognition and valuation of the contribution of its street trees to the continuing wellbeing of urban residents. From 2012, city authorities introduced a NYC Tree Valuation Method recognising NYC's urban forest as green infrastructure, integral to the health, beauty and vitality of the city and its residents. The tree canopy cleans the air, cools the streets, reduces storm water runoff, beautifies neighbourhoods and enhances property values.[327] NYC's Tree Valuation Method pays particular attention to the often neglected half of NYC's tree canopy estimated to be growing along streets and highways or on land reserved for open space and recreation, rather than in the city's various parks. Street trees are estimated to provide over $122 million each year in total benefits, with a benefit-to-cost ratio of 5:1.

New York City's Parks Department has created a database enabling the mapping of every street tree by a network of *TreesCount! 2015* volunteers, promoting a sense of ownership and stewardship.[328]

Recognition of the potential for multiple ecosystem service benefits generated by natural processes in urban areas is seen in restoration of the Mayes Brook as part of wider regeneration of Mayesbrook Park in East London[329] (Box 3.59) and also the River Quaggy in South London, UK[330,331] (Box 3.60). Both initiatives broke rivers out of engineered channels into meanders across regenerated urban parkland, delivering an ecosystem-based approach to urban flood management restoring riparian and floodplain wetland habitats valued for landscape aesthetics, local

---

[327] NYC Parks. (n.d.). *NYC Tree Valuation Method.* New York City Parks Department, New York. https://www.nycgovparks.org/pagefiles/128/New-York-City-Tree-Valuation-Method-05-04-2018__5b2ad0f011a85.pdf, accessed 13 January 2019.

[328] NYC Parks. (n.d.). *NYC's Street Trees.* New York City Parks, New York. https://tree-map.nycgovparks.org/, accessed 13 January 2019.

[329] Everard, M., Shuker, L. and Gurnell, A. (2011). *The Mayes Brook restoration in Mayesbrook Park, East London: An Ecosystem Services Assessment.* Environment Agency Evidence Report. Environment Agency, Bristol.

[330] Cowan, R., Hill, D. and Campbell, K. (2005). *Start with the Park: A Guide to Creating Sustainable Urban Green Spaces in Areas of Housing Growth and Renewal.* CABE Space, London.

[331] Potter, K. (2012). *Finding "Space for Water": Crossing Concrete Policy Thresholds in England.* In: Warner, J.F. and van Buuren, E.J. (Eds.), *Making Space for the River: Governance Experiences with Multifunctional River Flood Management in the US and Europe.* IWA Publishing, London, UK, pp. 89–102.

climate regulation, health benefits, access, education, wildlife and amenity.[332]

Box 3.59 Restoration of the Mayes Brook in Mayesbrook Park, East London

Restoration of the Mayes Brook and Mayesbrook Park, in the London Borough of Barking and Dagenham, was a multi-partner initiative from 2010/11. The project broke the Mayes Brook out of its former engineered channel behind security fencing, diverting it into newly-created meanders in a floodplain across the formerly neglected and underutilised Mayesbrook Park landscape (Fig. 3.28).

Fig. 3.26 A 'green wall' on the south face of the Edgware Road London Underground station cools the building and adjacent street, adds visual interest, hosts pollinators and buffers run-off from rainfall, amongst a wide range of beneficial ecosystem services in a densely developed urban setting. (Image © Dr Mark Everard)

---

[332] Everard, M. (2012). What have rivers ever done for us? Ecosystem services and river systems. In: Boon, P.J. and Raven, P.J. (Eds.), *River Conservation and Management*. Wiley, Chichester, pp. 313–324.

Project aims included delivering natural flood management, becoming the first park in the world adapting urban green space to climate change, contributing to biodiversity and river quality targets, and creating a long-term and sustainable asset in a socially deprived area of the city. Change in river ecology and the usage of the park by diverse groups of people are key indicators of the success of the scheme. Post-project assessment found that the restoration programme, taking account of likely quantifiable ecosystem service outcomes though acknowledging wider unquantifiable benefits, returned £7 of societal value for every £1 invested.[333]

Fig. 3.27   A roadside swale is one of many types of SuDS slowing flows of flood-water and promoting percolation to groundwater and natural purification processes. (Image © Dr Mark Everard)

[333] Everard, M., Shuker, L. and Gurnell, A. (2011). *The Mayes Brook Restoration in Mayesbrook Park, East London: An Ecosystem Services Assessment.* Environment Agency Evidence Report. Environment Agency, Bristol.

Box 3.60 Restoration of the River Quaggy in South London

Restoration of the River Quaggy in South London commenced as a multi-partner project in 1994 under a plan known as 'Operation Kingfisher'. Operation Kingfisher aimed at a complete river restoration of the urban River Quaggy from Chinbrook Meadows (Grove Park) to its confluence with the River Ravensbourne in Lewisham town centre. One of the key drivers of the project was the tendency of the Quaggy, then constrained in a heavily engineered channel, to flood neighbouring urban areas. The vision included a reconstructed floodplain in the parkland of Chinbrook Meadows. In 2002, some 300 metres of concrete channel through Chinbrook Meadows was bulldozed and the water diverted into reconstructed meanders across the park. Today, this restored reach of the River Quaggy flows across the park in meanders matching those in which it formerly flowed in the Nineteenth Century, rich in plant and animal life, revitalising a formerly uninteresting, little-used public recreational space.

Successes achieved at Chinbrook Meadows were followed in 2003 by two projects, also largely driven by the 'anchor service' of alleviation of flooding of homes and businesses, to release the Quaggy from an underground culvert in Sutcliffe Park. The River Quaggy project offers a range of linked environmental, educational and amenity benefits.[334]

A particular problem facing traditional high-density urban development is the creation of 'heat islands'. This term describes built-up areas that are hotter than nearby rural areas, generally due to lack of natural cooling functions and the trapping of static air, exacerbated by heat emissions from buildings and transport systems. The annual mean air temperature of a city with a population of 1 million people or more can be 1–3 °C warmer than its surroundings, or as much as 12°C warmer in the evening.[335] Heat islands affect communities in numerous ways, including increasing peak energy demand during the summer due to the use of air conditioning with associated greenhouse gas emissions, and contributing directly and indirectly to air pollution. Heat islands also have potentially significant implications for heat-related illnesses and premature deaths, with an estimated 47–95% increase in heat-related premature

---

[334] QWAG. (n.d.). *Restoring the River.* Quaggy Waterways Action Group (QWAG). https://qwag.org.uk/river-quaggy/restoring-the-river/, accessed 13 January 2019.

[335] US EPS. (n.d.). *Heat Island Effect.* US Environmental Protection Agency. https://www.epa.gov/heat-islands.

Fig. 3.28 The Mayes Brook was broken out of an engineered channel behind a security fence and diverted into newly-created meanders in a floodplain as part of the Mayesbrook Park restoration programme, delivering a wide range of societally beneficial ecosystem services. (Image © Dr Mark Everard)

deaths in New York City by 2050 under differing climate change scenarios relative to a 1990 baseline.[336] A study of the multiple beneficial ecosystem services provided by Sanjay Gandhi National Park (SGNP),[337] in the Indian state of Maharashtra, included observation of the Park's value as a

[336] Knowlton, K., Lynn, B., Goldberg, R.A., Rosenzweig, C., Hogrefe, C., Rosenthal, J.K. and Kinney, P.L. (2007). Projecting heat-related mortality impacts under a changing climate in the New York City Region. *American Journal of Public Health*, 97(11), pp. 2028–2034. 10.2105/AJPH.2006.102947.

[337] Sanjay Gandhi National Park and Wildlife and We Protection Foundation. (2019). *Developing Payment of Ecosystem Services Mechanisms for Sanjay Gandhi National Park—A Revenue Generating Model*. Sanjay Gandhi National Park and Wildlife and We Protection Foundation, Borivali, Mumbai.

Fig. 3.29 'Real time' temperature read-out installed at the Tokoroki Ravine in Tokyo, Japan, comparing the temperature between the valley and the surrounding city. (Image © Daisy Everard)

natural 'air conditioning' system for the surrounding megacity of Mumbai and Thane. The temperature within SGNP is lower by 3–5 °C than the temperature outside the Park at most times of the year. Amongst the wealth of ecosystem service benefits provided by green infrastructure and green spaces in urban environments, climate regulation at a local scale may be highly significant in the light of increasing global urbanisation and a changing climate. This benefit is appreciated in Japan, for example by a 'real time' digital read-out installed at the Tokoroki Ravine in Tokyo, comparing the temperature between the valley and the surrounding city (Fig. 3.29).

3.11   Economic Dimensions

The term 'values' has been used throughout the case studies in Chap. 3 of this book. Value is a plural concept, far broader than finance, alluding to the breadth of value systems. These include for trade, subsistence, regulatory and supporting processes maintaining the integrity and continued functioning of the natural world, spiritual, education and aesthetic dimensions, and more besides. It is important to think systemically about all of these values, both monetary and non-monetary. Otherwise, we simply perpetuate the unfortunate trajectory of market-driven appropriation of ecosystems, overlooking and as often undermining the many broader values they provide, now and into the future. It is this short-termism in the use of ecosystems that drives the cycles of socio-ecological degradation of the atmosphere, land and water underlying so many of today's sustainability concerns.

A key facet of the many prior regenerative case studies has been optimisation of outcomes across ecosystem services and their multiple linked beneficiaries, now and tomorrow. This wider vision of values needs to transform contemporary habits maximising one or a few near-term benefits by privileged sectors, with inevitable externalities borne by those without representation in today's decisions. When all potential benefits as well as disbenefits, many overlooked today, are taken into account, their cumulative value can be substantial. Equally, overlooking a spectrum of services in pursuit of maximising narrow, short-term returns can undermine net societal value, critically including the future resilience of the ecosystems providing those benefits as exemplified in Chap. 2, *Nature's sinking ark*.

Box 3.61 outlines some of the large cumulative values, represented in monetary terms, potentially accruing from ecosystems when multiple services are taken into account. None of these monetary values are absolute, but are merely representative in monetary terms of the scale of often otherwise incommensurable value. They also omit some values for which no valid representation could be made. However, they are at least indicative of the substantial scale of often underappreciated benefits flowing from ecosystems to society.

Box 3.61 Multiple Values Accruing from Ecosystems

An overall ecosystem service value was calculated for global forests at more than $16 trillion,[338] of which only 6% of temperate forest and 1.6% of tropical forest valuation is from extraction of 'raw materials' for which markets often provide primary management drivers.[339] Effectively, between 94% and 98.4% of their value is externalised from dominant market drivers.

Regeneration of coastal habitats for socio-environmental benefits is an increasingly mainstream policy priority due to assessments of their natural capital[340] and ecosystem services.[341] Rebuilding natural capital such as intertidal habitats can offer at least as great a return on investment as traditional engineered infrastructure.[342] The impetus for coastal restoration is then far from altruistic, as coastal habitats generate substantial societal value. The ecosystem services provided by the UK's coast were estimated as contributing £48 billion (2003 values), equivalent to 3.46% of Global National Income (GNI), though subject to major declines during the latter half of the twentieth century.[343] Globally, coastal resources are threatened by land reclamation for agriculture, port and other infrastructure, residential, industrial and other economic development, as well as poor management, pollution, changes to sediment flows and climate change including rising sea levels.[344] All of this undermines their benefits to society, with associated costs arising from ensuing damage and mitigation measures.

---

[338] Costanza, R., de Groot, R., Sutton, P., et al. (2014). Changes in the global value of ecosystem services. *Global Environmental Change*, 26, pp. 152–158.

[339] de Groot, R., Brander, L., van der Ploeg, S., et al. (2012). Global estimates of the value of ecosystems and their services in monetary terms. *Ecosystem Services*, 1, pp. 50–61.

[340] For example: Dorset Local Nature Partnership. (2016). *The Natural Place for Business: A Natural Capital Investment Strategy for Dorset*. Dorset Local Nature Partnership, Dorchester, 16pp. http://www.dorsetlnp.org.uk/hres/Dorset%20LNP%20Natural%20Capital%20Investment%20Strategy%202016.pdf, accessed 29 May 2016.

[341] Everard, M., Jones, L. and Watts, W. (2010). Have we neglected the societal importance of sand dunes? An ecosystem services perspective. *Aquatic Conservation: Marine and Freshwater Ecosystems*, 20, pp. 476–487.

[342] Natural Capital Committee. (2015). *Protecting and Improving Natural Capital for Prosperity and Wellbeing: Third 'State of Natural Capital' Report*. Natural Capital Committee, HM Government, London. https://www.naturalcapitalcommittee.org/, accessed 8 April 2015.

[343] Jones, L., Angus, A., Cooper, A., Doody, P., Everard, M., Garbutt, A., Gilchrist, P., Hansom, P., Nicholls, R., Pye, K., Ravenscroft, N., Rees, S., Rhind, P. and Whitehouse, A. (2011). *Chapter 11: Coastal Margins*. UK National Ecosystem Assessment, pp. 411–457. www.uknea.unep-wcmc.org/LinkClick.aspx?fileticket=9FfgOOAtUzU%3D&tabid=82, accessed 29 May 2016.

[344] UNEP. (2008). *An Overview of the State of the World's Fresh and Marine Waters*, 2nd ed. United Nations Environment Programme (UNEP). http://www.unep.org/dewa/vitalwater/article192.html, accessed 29 May 2016.

The World Bank[345] estimated that the global marine fish harvest could be 13% higher if fisheries were managed more sustainably. There is increasing evidence that up-front investments in fisheries recovery can generate substantial social and economic returns over time.[346]
Restoration of the Mayes Brook and Mayesbrook Park (outlined previously), taking account of likely quantifiable ecosystem service outcomes, was calculated to return £7 for every £1 invested.[347]
A study of the fringing mangroves of Mumbai concluded that six ecosystem services for which it was possible to transfer indicative economic values (fresh water, food, carbon sequestration, flood control, industrial and domestic wastewater treatment value, and recreation) conferred total benefits of over ₹Indian 500 million per year (US$ 7.73 million per year).[348]

Hard financial 'bottom line' values also underpin many of the schemes featuring as 'regenerative landscape' case studies. These include the SCaMP, Upstream Thinking and New York City water supply schemes, NFM in the Stroud Valley and elsewhere, aquifer recharge in Chennai, Arizona and other localities, the US CRP and nutrient management in Chesapeake Bay, various PES schemes such as in Costa Rica, New Zealand and under REDD+, and profitable pro-environmental farming in the UK and US. Unfortunately, contemporary statutory financial approval metrics still generally fail to account for the far wider basket of linked ecosystem service outcomes generated by these and other regenerative schemes.

Only if a broader conception of value than the blinkered perspective of contemporary market economics is taken into account, encompassing the benefits and costs of all interconnected ecosystem services, can a genuinely rounded economic appreciation be derived to inform sustainable policies governing wise ecosystem use and management decisions. Taking such an integrated approach, the Third Report (2015) annual report of

---

[345] The World Bank. (2015). *The Sunken Billions Revisited: Progress and Challenges in Global Marine Fisheries.*

[346] Holmes, L., Strauss, C.K., de Vos, K. and Bonzo, K. (2014). *Towards Investment in Sustainable Fisheries: A Framework for Financing the Transition.* Environmental Defence Fund, New York; The Prince of Wales' International Sustainability Unit, London; 50in10, Washington, DC.

[347] Everard, M., Shuker, L. and Gurnell, A. (2011). *The Mayes Brook Restoration in Mayesbrook Park, East London: An Ecosystem Services Assessment.* Environment Agency Evidence Report. Environment Agency, Bristol.

[348] Everard, M., Jha, R.R.S. and Russell, S. (2014). The benefits of fringing mangrove systems to Mumbai. *Aquatic Conservation: Marine and Freshwater Ecosystems*, 24, pp. 256–274.

the UK's Natural Capital Committee (NCC), the 'State of Natural Capital' report *Protecting and Improving Natural Capital for Prosperity and Wellbeing,*[349] recognised 'natural capital deficits' built up over the long term that are proving costly to societal wellbeing and the economy. To counter this degrading trend, the NCC incorporated a 25-year plan for restoration of nature capital into its recommendations to UK government. This recommendation was subsequently at least in part reflected in the 2017 *A Green Future: Our 25 Year Plan to Improve the Environment*[350] UK government strategy. The 25-year Plan includes, as one of three strategic themes, a strong economic case for investment in the creation and restoration of several habitat types (see Box 3.62).

Box 3.62 The NCC 25-Year Case for Investment in Habitat Recreation

The UK's Natural Capital Committee (NCC) identified habitat creation as a net value-add for UK society. Strong economic cases were presented for creating up to 250,000 additional hectares of woodland, restoring 140,000 hectares of upland peatland, and creating around 100,000 hectares of wetland. Location is of fundamental importance in terms of the production, access to and benefits realised from the ecosystem services that they would produce.

- Wetland creation, for example, delivers optimal value when located in areas of suitable hydrology, upstream of major towns and cities where flood regulation benefits and public access are maximised, and avoiding areas that would displace high-grade agricultural land.
- The NCC study also made a strong economic case for commercial fish stock restoration, particularly of white fish (such as cod) and shellfish for which populations today are considerably below optimal levels.
- Intertidal habitat creation could also meet objectives set out in Shoreline Management Plans, returning substantial net societal benefit.
- Creation of urban greenspaces was assessed as likely to provide a high return on investment through provision of multiple ecosystem services.

Reflecting the mainstream importance of such societal investments, the NCC found that returns on investment from this rebuilding of natural capital were at least as great as those from investment in traditional engineered infrastructure.

---

[349] Natural Capital Committee. (2015). *Protecting and Improving Natural Capital for Prosperity and Wellbeing: Third 'State of Natural Capital' Report.* Natural Capital Committee, HM Government, London. https://www.naturalcapitalcommittee.org/, accessed 8 April 2015.

[350] HM Government. (2018). *A Green Future: Our 25 Year Plan to Improve the Environment.* HM Government, London.

The term 'natural capital', has been used variously throughout this book, and the concept of 'natural capital' has gained political support over recent years. The World Forum on Natural Capital[351] defines natural capital as "*...the world's stocks of natural assets which include geology, soil, air, water and all living things*", and the term has undoubtedly been helpful in improving business and government awareness of the benefits and risks associated with management of natural assets. However, a word of caution is required. Business and government are comfortable with the concept of capital, which is seen in potentially owned property such as road and rail networks, pipework, fisheries and forests. However, what is of fundamental importance is not the stock of capital, but the flows of benefits that stem from it in the form of primarily publicly beneficial ecosystem services. A field, for example, can be bought, sold and used for farming as a capital asset, but the characteristics and stewardship of the land will determine its flows of linked benefits or disbenefits in terms, for example, of providing habitat for wildlife, hydrological buffering of floods and droughts, aesthetic and recreational opportunity, carbon sequestration and nutrient cycling. The diverse beneficiaries of these public services have no ownership of the stock, and limited influence on the policy environment governing its uses by generally private owners. Conflating stocks and flows, or natural capital and ecosystem services, has risks of market capture that need to be clearly understood and managed.

3.12   Corporate Leadership

Global businesses are awakening to the benefits of sustainability in terms of improved investment prospects and the long-term corporate viability. The NGO Natural Capitalism Solutions summarises conclusions of an overview of many leading companies as "*When those wild-eyed environmentalists at Goldman Sachs tell you that the companies that are the leaders in sustainable, social and good governance policies have*

---

[351] https://naturalcapitalforum.com/about/.

25% higher stock value than their less sustainable competitors, there's a business case for behaving in ways that are more responsible to the planet and to people".[352]

The value of natural systems and processes is not lost on many businesses, and particularly those most directly dependent on primary ecosystem resources. We have seen this, for example, in Upstream Thinking, SCaMP and other water service-driven schemes now working more closely with measures to protect raw water quality at source, rather than investing more heavily in downstream purification of abstracted water carrying a higher pollutant load. Other businesses too have engaged in landscape management to regenerate water resources in catchments where they operate, offsetting their demands and benefitting local interests as a contribution to their societal 'licence to operate' (Box 3.63). Although corporate claims are contested by a number of NGOs, the bold aspirations of these beverage giants demonstrate practical responses that the business sector can make, self-beneficially whilst also engaging with global sustainability challenges.

Box 3.63 Corporate Leadership in Landscape Management for Water Security

Diageo, a global leader in beverage alcohol, launched a 'Water Blueprint'[353] in December 2014. This was part of an ambitious sustainable development strategy including better management of carbon emissions, cutting water use in half, returning waste water to the environment safely across supply chains and, significantly, to replenish water resources in water-stressed areas to an equivalent quantity used in final products. This is enacted through projects such as reforestation, wetland recovery and support for improved farming techniques.[354]

---

[352] Natural Capitalism. (2013). *Sustainability Pays: Studies that Prove the Business Case for Sustainability.*

[353] Diageo. (2015). Diageo Water Blueprint. http://www.diageo.com/en-row/NewsMedia/Pages/resource.aspx?resourceid=2730, accessed 7 September 2015.

[354] Diageo. (2014). *Diageo Commits to New Ambitious Sustainability and Responsibility Targets for 2020.* http://www.diageo.com/en-row/newsmedia/pages/resource.aspx?resourceid=2411, accessed 8 August 2015.

SABMiller, a multinational brewing and beverage company, has helped fund four water recharge dams located on natural fissures in Rajasthan, India, contributing to regeneration of groundwater resources depleted by its operations.[355] SABMiller is amongst a group of major water-consumptive industries seeking to replicate traditional groundwater infiltration technologies at industrial scale to replenish resources to match factory demand in the arid and semi-arid Indian state of Rajasthan.[356]

In 2009, PepsiCo India claimed that it had become the first business to achieve 'Positive Water Balance' in the beverage world through several initiatives to replenish water in communities, including facilitating construction of check dams.[357]

In August 2015, Coca Cola announced that it replenished an estimated 94% of 2014 sales, on track to meeting its 100% water replenishment goal.[358] The company reported in 2018 that it continues to meet its goal to *"By 2020, safely return to communities and nature an amount of water equal to what we use in our finished beverages"*.[359]

Other corporates with a high dependence on primary natural resources have also engaged actively with multi-stakeholder approaches to stewardship, as a mechanism to maintain continued flows of essential capital for their businesses into the future. A wide and expanding range of market-differentiating certification schemes has been led by businesses, or consortia with significant business representation and engagement. The Forest Stewardship Council (FSC) (Box 3.64), Marine Stewardship

[355] Balch, O. (2012, October). Mouth watering. *Green Futures Special Edition—Water Works: Green Solutions for a Blue Planet*, pp. 22–23.

[356] Confederation of Indian Industry. (n.d.). *Breaking the Boundaries in Water Management: A Case Study Compendium*. Confederation of Indian Industry Northern Region, Jaipur.

[357] Pepsico India. (n.d.). *Conserving the World's Most Precious Asset: Water*. Pepsico India. http://www.pepsicoindia.co.in/purpose/environmental-sustainability/replenishing-water.html, accessed 6 June 2016.

[358] Coca-Cola. (2015). *Coca-Cola on Track to Meet 100% Water Replenishment Goal*. Coca-Cola, 25 August 2015. http://www.coca-cola.co.uk/newsroom/press-releases/coca-cola-on-track-to-meet-100-percent-water-replenishment-goal, accessed 6 June 2016.

[359] The Coca Cola Company. (2018). *Collaborating to Replenish the Water We Use*. The Coca-Cola Company, 29 August 2018. https://www.coca-colacompany.com/stories/collaborating-to-replenish-the-water-we-use, accessed 21 February 2019.

Council (MSC) (Box 3.65), Rainforest Alliance (Box 3.66) and the Roundtable on Sustainable Palm Oil (Box 3.67) exemplify market-based schemes in different states of maturity, credibility and market impact.

Box 3.64 The Forest Stewardship Council (FSC) Scheme

The Forest Stewardship Council[360] (FSC) is a global forest certification system covering the two components of Forest Management and Chain of Custody certification. These allow consumers to identify, purchase and use wood, paper and other forest products with an audited chain of custody from well-managed forests and/or recycled materials.

The FSC was set up in 1994 by a consortium of interests that, at the time, seemed highly improbable. This consortium ranged from environmental NGOs (WWF, Friends of the Earth, Greenpeace), indigenous forest dwellers, professional forestry interests and major retailers such as Sweden's IKEA and the UK's Kingfisher Group (notably owning the DIY chain B&Q). All shared a desire for a workable system that would promote responsible forest management practices and create a clear market for them. This bold, ground-breaking leadership created a scheme that was not only well in advance of what national government or intergovernmental groups had achieved, but seemingly also what they could at the time conceive.

The business drive was essential to make the FSC scheme workable. As of December 2018, 1,606 FSC forest certificates had been issued in forests in 85 countries spanning 200,963,183 hectares (over twice the land area of Portugal), with 35,772 'chain of custody' certificates issued in 123 countries.[361]

Box 3.65 The Marine Stewardship Council (MSC) Scheme

The Marine Stewardship Council[362] (MSC) scheme was co-founded in 1996 by Unilever and the WWF, modelled on the successes and processes of the FSC. Clear business benefits from sustainable fish stocks accrue to Unilever, one of the world's largest manufacturers of food and household goods estimated to purchase around 20% of Europe's fish catches.

---

[360] https://www.fsc-uk.org/en-uk.

[361] FSC. (2019). *FSC: Facts & Figures, December 3, 2018*. Forest Stewardship Council. https://fsc.org/en/page/facts-figures.

[362] https://www.msc.org/.

The goal of the MSC is to achieve solutions to the problem of declining fish stocks. Sustainability is the key criterion, simultaneously promoting environmental and economic benefits from maintaining the biodiversity and ecological processes of the marine environment. The MSC was constituted as an independent, global, not-for-profit body in 1999, with independent auditing and certification to offer a similar transparent 'chain of custody' to that of the FSC for the transfer of marine products from fishery to plate. In its *Global Impacts Report 2017*, the MSC recorded that *"12% of global marine wild catch is certified to the MSC Fisheries Standard, the market for certified sustainable and labelled seafood is worth over US$5 billion and the program is widely recognised as the most rigorous and credible indicator of environmental sustainability and traceability in the seafood sector"*.[363]

Box 3.66 The Rainforest Alliance Scheme

Rainforest Alliance[364] is another example of a certification scheme for consumer products, produced by means that conserve biodiversity and ensure sustainable livelihoods by transforming land-use practices, business practices and consumer behaviour.

The Rainforest Alliance, incorporated in 1987 as an international non-profit organization, constitutes a growing network of farmers, foresters, communities, scientists, governments, environmentalists, and businesses dedicated to conserving biodiversity and ensuring sustainable livelihoods.

Box 3.67 The Roundtable on Sustainable Palm Oil (RSPO) Scheme

The Roundtable on Sustainable Palm Oil[365] (RSPO) is a not-for-profit organisation comprising stakeholders from seven sectors of the palm oil industry: oil palm producers, processors, traders, consumer goods manufacturers, retailers, banks/investors, and environmental and civil non-governmental organisations (NGOs). The RSPO includes over 3,000 members worldwide.

---

[363] MSC. (2017). *Marine Stewardship Council: Global Impacts Report 2017*. Marine Stewardship Council, London. https://www.msc.org/docs/default-source/default-document-library/what-we-are-doing/global-impact-reports/msc-global-impacts-report-2017-interactive.pdf.

[364] https://www.rainforest-alliance.org/.

[365] https://rspo.org/about.

> The RSPO aims to develop and implement global standards for sustainable palm oil, intending to transform markets to make sustainable palm oil the norm. Palm oil is in many everyday consumer products, though its production today is by large majority beset with a wide range of highly damaging environmental and social outcomes. The RSPO has developed a set of environmental and social criteria with which member companies are required to comply in order to attain Certified Sustainable Palm Oil (CSPO) status. When they are properly applied, these criteria can help to minimise the negative impact of palm oil cultivation on the environment and communities in palm oil-producing regions.
>
> The RSPO is at a far earlier stage of evolution than the FSC and the MSC, but is at least a start on a long journey toward the aspiration of sustainable sourcing of palm oil.

There are also inspiring examples of corporate leadership in driving public awareness of issues pertaining to the crucial importance of ecosystems for continuing human wellbeing. One such influential and impactful case is that of Penny Market, a budget German supermarket, highlighting to its customer base the vital importance of the pollination services provided by bees (Box 3.68).

Box 3.68 Corporate Promotion of Societal Awareness of the Importance of Bees in Germany's Penny Market

The German Penny Market discount supermarket store in Hannover surprised its substantial customer base on Monday 14th May 2018 by substantially emptying its shelves, removing products pollinated by bees.[366] The purpose of this was to draw public attention to and stimulate concern about the significant implications for consumers if bees died out.

Apples, zucchini, baked goods, chocolate, sweets coated with beeswax, some marinated meats and chamomile-scented toilet paper were just some of the many products taken off the shelves. In all, Penny Market assessed that approximately 60% of its 2,500 products are directly or indirectly

---

[366] The Local. (2019). Bee-n and gone: Hanover supermarket warns customers of bee-less world. *The Local*, 15 May 2018. https://www.thelocal.de/20180515/hanover-shop-empties-shelves-of-bee-pollinated-products, accessed 11th May 2019.

dependent on bee pollination. Pollination services globally have substantial economic value, though their central roles in the continued functioning of global ecosystems is beyond monetary representation. Bee populations have been in drastic decline over recent decades, with frightening potential consequences.

Penny Market had not informed its customers in advance about this symbolic action, a company spokesman stating that *"We were hoping for a eureka moment"*. The supermarket's campaign preceded the United Nations' first ever World Bee Day on 20 May 2018.

It is not, however, just in the sourcing of materials and energy that businesses interact with ecosystems; interdependencies with socio-ecological systems are far more profoundly integrated than that. The manufacturing, transportation and distribution, waste generation and all other elements of business metabolism are, whether recognised as such or not, as highly interactive with and influential upon ecosystems and people. 'Clean production' is preventive initiative intended to minimize waste and emissions and maximize product output, and 'Corporate Social Responsibility' (CSR) is a self-regulating and generally transparently reported business model that helps a company to be socially accountable. These and many more initiatives besides, beyond the scope of this book's primary focus on ecosystems, are increasingly embraced with the dawning of awareness of how the functions of enterprises mesh with and ultimately depend upon supporting social and natural environments.

But it is not just the performance of a business that has a socio-ecological life cycle: so too do its products. In fact, elements of the life cycle of a product once it leaves the factory gate can have as great, or potentially greater, impact on supporting ecosystems as the sourcing of raw materials or production processes. The mass environmental accumulation of plastic litter is one example of folly borne of the lack of foresight about the fate of products at end-of-life. Though graphic, growing plastic pollution is in reality just the tip of the metaphorical iceberg of our collective blindness to impacts on supporting ecosystems in the design and acceptability of products framed purely on immediate, short-term utility. So too, we have been naïve about inputs during the useful life of products—energy, cleaning chemicals and other maintenance demands, water, preservatives such as routine painting and anti-fouling, and so forth—that may be highly damaging to supporting ecosystems even if the raw material sourcing and manufacturing phases of the life cycle were responsibly managed.

These wider aspects of the total socio-ecological life cycle of businesses, product design, use and fate are somewhat beyond the scope of this book, though nonetheless highly germane to closer integration with the processes and sustainability of supporting ecosystems. But one interesting initiative—biomimicry, or "design inspired by nature"—is of interest in its encapsulation of the intent to align design and production processes far more closely with nature's solutions and processes (Box 3.69).

Box 3.69 Biomimicry: Design Inspired by Nature

The term 'biomimicry' defines a pathway of innovation seeking sustainable solutions to societal challenges through emulation of patterns and strategies found in natural systems resulting from long evolution.[367] The ultimate goal is to develop products, processes and policies that work in synergy with ecosystem processes. Through billions of years of evolution, entailing both symbiotic solutions as well as the extinction of less successful adaptations, the natural world has solved many of the problems that humans are grappling with today.

Whereas society has used 'heat, beat and treat' methods (often at high temperature, with intensive mechanical processes and with potentially harsh chemicals), nature produces such high-performance solutions a spider silk, photosynthetic processes, natural medicines and forest ecosystems at ambient temperatures, in water and without toxic effluent. Biomimicry takes inspiration for mechanical structures, chemical solutions and production processes from solutions evolved in the natural world as frameworks to develop more sustainable products and technologies.

---

[367] Benyus, J.M. (1998). *Biomimicry, Innovation Inspired by Nature*. William Morrow/Quill.

3.13   Devoting Half the World to Nature

The natural world and its diverse forms, functions and services is the primary capital upon which humanity's wellbeing depends. Though we live in a modern world of great technological sophistication, by majority we remain culturally ignorant about the impacts of our activities and methods on the natural world, and our vastly impoverished and rapidly declining capital of supportive ecosystems. A key challenge is to halt the pace of destruction if we are to support the needs of a growing human population. However, this alone is insufficient; we must with urgency go substantially beyond this, 'rebuilding the Earth' by restoring the primary natural capital upon which continuing security and opportunity depends.

Rebuilding ecosystems at wide and connected scales is a major challenge if human potential is not to be limited by ecosystem capacity. We need only look at habitats such as polar regions, high mountains and deserts to see how limited natural carrying capacities restrict human numbers and options; we are incrementally imposing the same types of limitations on all habitats globally as we allow ecosystems to continue to degrade.

The challenge of allowing the regeneration of natural systems and processes is daunting in the face of current degrading trends, legacy policy and technology assumptions, and vested interests entrenched in the market system. However, the wealth of exemplars in this chapter—from right across the globe, from local to large landscape scale, and spanning nations in varying stages of development—shows that a regenerative approach is not only possible but is occurring albeit on a fragmented basis. These initiatives also cumulatively highlight how wider shifts in societal responses, backed up by a concerted and informed revision of policies and actions, can serve to simultaneously safeguard or regenerate ecosystem vitality and integrally linked human prospects. This is the case from the massive scale of China's Loess Plateau and the Ethiopian Highlands, to catchment-scale activities such as regeneration of rivers in Alwar District or management of flooding in Gloucestershire, and right down to day-to-day transitions in as prosaic an activity as recreational fishery management (see Box 3.70).

Box 3.70 Transitions in Recreational Fishery Management

Fishery management for recreational angling is increasingly taking ecosystem-based consideration of the needs of fish populations (Fig. 3.30). Rivers, lakes and estuaries need to contain enough water of adequate quality, with food to support mixed fish species, and a diversity of habitat for fish breeding, feeding and refuge needs throughout their various life stages.[368] Management informed by that understanding is also vital for influencing the wider environment of policy and practice to address the many pressures on aquatic environments, of which fishery uses are one amongst many interlinked societal benefits.

This contrasts with the caricature of the shotgun-wielding river fishery manager with a slash-and-burn approach to riparian habitat, restocking fish and culling predators to maintain a standing crop of accessible fish. This parody was perhaps warranted in a less enlightened period after the Second World War. However, this paradigm has, with some exceptions, been substantially reversed over successive decades, at least across Europe. Today, progressive management of river fisheries seeks to identify and protect, and to restore where degraded, habitat important for the life cycles of fishes, using measures such as creating riparian buffer zones (addressed in consideration of progressive farming practices), flushing silt from spawning gravels, retaining or installing large woody matter into streams to host food and to diversify flow regimes and physical habitat, and opening up 'fry bays' in habitat-poor river margins.[369,370]

Documented angling catch returns are also increasingly important as budgets for statutory monitoring of fish populations are cut back, or where routine environmental monitoring does not occur, providing evidence of trends in fish populations indicative of broader ecosystem health.[371] Angling also serves multiple purposes including connecting people to the natural environment, encouraging socialisation across cultural, age, socio-economic and other barriers, and providing revenue for river protection. Fishery interests also mobilise significant political power, wielded though a range of NGOs to influence policy and practice better to safeguard aquatic environments.

In developing countries, the economic value of recreational angling activities can be an influential force for conservation of river ecosystems

---

[368] Everard, M. and Pinder, A.C. (2018). Angling for sustainable fishing. *Environmental Scientist*, December 2018, pp. 29–25.

[369] Everard, M. (2015). *River Habitats for Coarse Fish: How Fish Use Rivers and How We Can Help Them!* Old Pond Publishing, Sheffield.

[370] Wild Trout Trust. (2017). *The Wild Trout Survival Guide*, 4th ed. Wild Trout Trust, Waterlooville.

[371] Environment Agency. (2018). *A Survey of Freshwater Angling in England—Phase 1: Angling Activity, Expenditure and Economic Impact*. Environment Agency, Bristol.

where revenues from visiting recreational anglers can provide powerful economic incentives for local communities to hold back from destructive fishing practices (dynamiting, pesticides, non-selective gill nets), and instead to self-police rivers where live fish have a greater value to them than dead fish.[372]

Metrics such as viable fish populations or other measures of biodiversity have to be understood in an integrated way as indicators of ecosystem viability with profound linked implications for human well-being. As one example of the weakness of recognition of the integrative role of such indicators, the UK government adopted a sustainable development indicator for Atlantic salmon stocks but, after serially

Fig. 3.30 Transitions in recreational fishery management in Europe are focusing increasingly on habitats and ecosystem processes supporting self-sustaining fish populations, also delivering a range of linked ecosystem service benefits. (Image © Dr Mark Everard)

---

[372] Everard, M. and Kataria, G. (2011). Recreational angling markets to advance the conservation of a reach of the Western Ramganga River, India. *Aquatic Conservation*, 21(1), pp. 101–108.

failing to meet an unambitious target, the indicator was subsequently dropped rather than seen as a call to action (Box 3.71). The significance of the indicator is entirely lost if regarded as merely desirable, or for altruistic purposes unrelated to the viability of ecosystems supporting human security and progress. However, when informed by vision and cross-policy integration, restoration of ecosystem functions can not only prove beneficial to biodiversity but also to the natural functions serving multiple societal needs, now and tomorrow. When viewed in a connected way, they add urgency and motivation for appropriate cross-policy action to safeguard critical natural capital and ecosystem services underpinning a diversity of human needs.

Box 3.71 The UK Government Indicator for Sustainable Salmon Stocks

A set of sustainable development indicators set by UK government in the late 1990s included a Conservation Limit for Atlantic salmon (*Salmo salar*) in British rivers. Conservation Limits reflect stock size below which further reductions in spawning numbers are likely to result in significant reductions in the numbers of juvenile fish produced in the next generation, and hence returning adults from that year class. Salmon Conservation Limits are set specifically for river systems, recognising the genetically distinct stock in each river and the need to maintain a sustainable population to retain genetic diversity. Atlantic salmon are both a priority species (scheduled under the EU Habitats Directive) and an indicator of ecosystem health.

Government set an unambitious target of 27 qualifying rivers, out of 40 rivers assessed annually. In reality, all 40 rivers should be meeting their Conservation Limit targets if there is serious commitment to attaining a sustainable future, supported by critical aquatic and other environmental resources and reversing the ongoing degrading cycles from which human opportunity can only become increasingly limited. Between 1997 and 2007, the number of rivers with sustainable salmon stocks varied between 13 and 25, rising to 28 in 2008 and declining to 16 in 2009.

Various metrics of river quality, bird populations and sustainable managed marine fisheries remained in the *Sustainable Development Indicators July 2013*[373] report and the *Sustainable Development Indicators, July 2014*

---

[373] Defra. (2013). *Sustainable Development Indicators, July 2013*. Department of Environment, Food and Rural Affairs, London. https://www.gov.uk/government/uploads/system/uploads/attachment_data/file/223992/0_SDIs_final__2_.pdf.

report.[374] However, the salmon indicator had been dropped. Serial failure to meet the salmon-related indicator would not have told a rosy political story.

In fact, progress against the various indicators providing an overview of progress toward a sustainable economy, society and environment in the 2014 report highlight that only 18 out of 60 measures showed poor overall progress with improvement over the long-term, with 27 showing improvement over the short-term. 9 out of 60 measures showed deterioration over the long-term, with 10 showing deterioration over the short-term.

In conservation discourse, there is debate about a 'land sharing' as opposed to a 'land sparing' approach. In essence, land sharing entails combining biodiversity conservation with food production, generally at lower intensity and yield, on the same parcels of land. Agri-environmental policies and incentives promoted by the European Union, wherein compensation is offered for loss of production for pro-environmental measures, is a continental-scale example of the land sharing approach. Rewilding at Knepp Estate offers a further, more extreme example. By contrast, land sparing segregates areas protected for wildlife from those devoted to more intensive agricultural production and other forms of development. A land sparing approach entailing delineation of legally protected conservation areas to protect biodiversity and associated ecosystem services is generally favoured in the USA, where lower average population densities compared to Europe enable scheduling of large 'wilderness areas' and National Parks separated from extensive areas of intensive agriculture. However, global evidence of steep declines in biodiversity suggests that neither the sharing nor sparing philosophies represent a panacea in the face of growing human population and resource demands, with nearly 40% of the earth's terrestrial surface already converted for agriculture. Outcomes from grey wolf and beaver reintroductions and otter recovery also highlight that, even in extensive 'spared' landscapes, there is plenty of scope for further regeneration of ecosystem processes with significant gains in wildlife and ecosystem services.

---

[374] Office for National Statistics. (2014). *Sustainable Development Indicators, July 2014*. Office for National Statistics, London. https://webarchive.nationalarchives.gov.uk/20160105183323/http://www.ons.gov.uk/ons/rel/wellbeing/sustainable-development-indicators/july-2014/sustainable-development-indicators.html.

There is consequently a need for a more nuanced approach. In Bhutan, where pro-conservation beliefs and initiatives are deeply culturally embedded, more than half (51.32%) of the territory is designated as National Parks and wildlife reserves, with 9% of land area scheduled as corridors linking protected areas constituting the Bhutan Biological Conservation Complex (B2C2).[375] In Japan, approximately 25 million hectares of land area is forested, covering some 67% of the country (over twice the world's average of 29% forest cover), albeit that this resolves down to a small per capita value of 0.2 hectares per capita due to the country's large human population. These high forest area values in Bhutan and Japan are as much a feature of the mountainous landscape, largely unsuitable for agricultural and other forms of urban and industrial development, emphasising that different approaches to sparing and sharing can be creatively adapted to topography, population density and other context-sensitive factors, with more effective outcomes than imposition of uniform, inflexible policy. A study in the USA found that, notwithstanding the conservation benefits of land sparing, a land sharing approach in more complex landscapes was effective as the majority of plant species in agroecosystems were found in small fragments of non-crop habitat.[376] The study concluded that, in landscapes with little non-crop habitat, botanical richness can be more readily conserved through land-sparing approaches at a highly localised scale, such as field margins and corners. This conclusion breaks down the coarse distinction between land sparing and land sharing by taking a more locally nuanced approach, informed by local-scale landscape characteristics and opportunities. The interdependence of agriculture and biodiversity is complex and not always well understood.[377] However, it is clear that local intelligence is essential for optimising protection of nature and natural processes in farmed and otherwise developed landscapes. This

---

[375] WWF. (n.d.). Conservation in Bhutan. *WWF.* http://www.wwfbhutan.org.bt/projects_/.

[376] Egan, J.F. and Mortensen, D.A. (2012). A comparison of land-sharing and land-sparing strategies for plant richness conservation in agricultural landscapes. *Ecological Applications*, 22(2), pp. 459–471.

[377] Tscharntke, T., Clough, Y., Wanger, T.C., Jackson, L., Motzke, I., Perfecto, I., Vandermeer, J. and Whitbread A. (2012). Global food security, biodiversity conservation and the future of agricultural intensification. *Biological Conservation*, 151, pp. 53–59. https://doi.org/10.1016/j.biocon.2012.01.068.

need not be in opposition to appropriate top-level policies, which ideally should be enabling rather than inflexibly prescriptive.

Evolutionary biologist E.O. Wilson proposed a radical 'Half Earth' goal wherein ecosystems spanning half the Earth are reserved for nature to retain ecological viability and the capacities of ecosystems to sustain our needs into the future.[378] This philosophical approach does not necessarily entail abandoning human occupation and uses, but incorporates chains of uninterrupted corridors linking biodiversity reserves that enable wildlife to move south-to-north to accommodate climate warming and east-to-west responding to changes in rainfall. In effect, the Bhutan Biological Conservation Complex (B2C2) is already meeting this goal at a national scale, albeit in a land of extreme Himalayan topography and low population density. Linear habitat such as roadside verges, railways, river corridors and walking/cycling routes can potentially be diversified to improve their already significant roles as wildlife corridors.[379]

Of course, abandoning half the countries on Earth is also a non-starter politically, as well as serving to surrender the distinctive ecosystems of the populated half to a doubling of human pressures. But the progressive embedding of nature and ecosystem processes into many areas of human activity—water management, farming and city ecosystems, urban and wilder forests, and incorporation in business-led stewardship schemes—can make space for regeneration of the natural, restored and emulated ecosystems providing ecosystem services vital to reopening human opportunity and security. The previous example of potential establishment of a strategic network of marine protected areas (MPAs) in areas important for a range of natural processes to protect or allow restoration of marine ecosystems in the wider ocean, and the many and valuable benefits they confer upon humanity, is one example of how this might be achieved. We are not 'surrendering' land or aquatic systems for nature, as some might conceive it, but investing throughout societal endeavours in the most vital capital underpinning our collective global wellbeing and future.

---

[378] Wilson, E.O. (2013). *The Social Conquest of Earth*. Liveright Publishing Corporation, New York.
[379] Spellerberg, I.F. and Gaywood, M.J. (1993). *Linear Features: Linear Habitats and Wildlife Corridors*. English Nature Research Reports No. 60. English Nature, Peterborough.

4

Our Conjoined Future

Since publication in 1987 of *Our Common Future*, the report of the World Commission on Environment and Development containing the famous 'Brundtland definition', sustainable development has entered public and political discourse. This definition, "*…development that meets the needs of the present without compromising the ability of future generations to meet their own needs*", is a visionary and explicitly intergenerational commitment. However, much subsequent regulatory transposition has reframed the concept in far weaker terms, more or less reducing it to an incremental 'lightening of the footprint' of societal activities on ecosystems. This profound weakening has consequently eroded understanding and urgency in the public mind. Not only does it implicitly assume that the supportive capacities of ecosystems are static, when in reality we know them to be in steep decline, but it also overlooks their already substantially degraded state and, with it, diminished prospects for human security and opportunity.

It is consequently essential for us to elevate our collective vision of sustainable development to one of rebuilding degraded natural systems as crucial underpinnings of a decent quality of life and future for humanity,

© The Author(s) 2020
M. Everard, *Rebuilding the Earth*, https://doi.org/10.1007/978-3-030-33024-8_4

free from conflict over dwindling resources.[1,2] As the fate of humanity is indivisibly tied to that of the ecosystems that support it, investment in a regenerative approach to landscape and natural resource uses offers the only viable bedrock for sustainability.

4.1   Systemic Thinking and Action

So how do we approach 'Rebuilding the Earth', learning from the foundations of global exemplars? First and foremost, it is vital that we recognise that our interdependence with ecosystems is utterly vital to all dimensions of human needs. The preceding overviews highlight how humanity and nature are integrally interconnected in tightly linked socio-ecological systems, be this relationship degenerative or regenerative. Today, we are seeing a prevalence of degenerative socio-ecological cycles, but are also blessed with a broad palette of examples of regenerative practices wherein proactive measures are once again regenerating the ecosystems providing resources securing and supporting human potential. The knowledge and tools are in our hands, though putting that reality into practice to reform a fragmented formal and informal policy environment is a challenge society has yet to fulfil. As for any damaging addiction—ours today to short-term economic profit-taking including vested interests disregarding longer-term consequences—recognition of the problem is the first step.

A strategic priority is the recognition that today's challenges are not unidimensional, with simple objective criteria defining indisputable public good or equity. Instead, they are what is known as 'wicked problems', difficult or impossible to solve because of incomplete, contradictory and changing requirements and complex interdependencies.[3] Every challenge is embedded within a complex socio-ecological system. It is in the nature of systems that all elements are fully interactive. Therefore, any use or management activity affecting a resource ramifies across the whole ecosystem and its diverse beneficiaries (or victims). The components of socio-

---

[1] Everard, M. (2013). *The Hydropolitics of Dams: Engineering or Ecosystems?* Zed Books, London.
[2] Everard, M. and Longhurst, J.W.S. (2018). Reasserting the primacy of human needs to reclaim the 'lost half' of sustainable development. *Science of the Total Environment*, 621, pp. 1243–1254.
[3] Rittel, H.W.J. and Webber, M.M. (1973). Dilemmas in a general theory of planning. *Policy Sciences*, 4, pp. 155–169.

ecological systems are as interdependent as notes in music, letters in a word, words in a sentence or codons of DNA, all of which lack meaning outside of the system of which they are components. Ecosystems comprise complex systems, defined as "…*a dynamic complex of plant, animal and micro-organism communities and their non-living environment interacting as a functional unit*",[4] their properties arising from dynamic interactions between living and abiotic constituents. Just as an amino acid in isolation lacks any vestige of the enzymatic, hormonal, structural or other property of the protein of which it is part, but can disrupt that function if missing or impaired in any way, a narrow use or management response in a complex socio-ecological system can disrupt the workings of the whole.

Society has, as we have seen, tended historically to manage ecosystems to maximise narrow outputs—food, water, forest timber, mined products and so forth—rewarded by markets for this blinkered framing with scant awareness of the wider interconnected suite of ecosystem services and their associated beneficiaries. Worse still, this presumption in favour of immediate profit-taking is locked into societal habits including in the ways that markets, investments and property rights are constructed, as well as legacy regulations that establish societal expectations and norms. However, neither natural nor managed ecosystems deliver single services in isolation. Instead, they generate linked sets of ecosystem services often referred to as 'environmental services'[5] or 'bundles'.[6] Optimising production or maintenance of these linked services demands a more integrated basis for decision-making, in place of the habitual short-term horizon and maximisation of narrowly framed outputs.

So how best to challenge and redress this myopia, so deeply entrenched in our markets, legacy assumptions, technologies and legislation? Two key concepts, introduced previously in this book, provide keys to unpick the lock. The first is to regard drivers for ecosystem management or uses—food production, recreation, flood management or other—not as

---

[4] Convention on Biological Diversity. (n.d.). *Description*. Convention on Biological Diversity. https://www.cbd.int/ecosystem/description.shtml, accessed 11 November 2016.

[5] Schomers, S. and Matzdorf, B. (2013). Payments for ecosystem services: A review and comparison of developing and industrialized countries. *Ecosystem Services*, 6, pp. 16–30. https://doi.org/10.1016/j.ecoser.2013.01.002i.

[6] Balvanera, P., Quijas, S., Martín-López, B., et al. (2016). The links between biodiversity and ecosystem services. In: Potschin, M., Haines-Young, R., Fish, R. and Turner, R.K. (Eds), *Routledge Handbook of Ecosystem Services*. Routledge, London, pp. 45–61.

a sole priorities but as 'anchor services' around which management options and exploitation techniques can be considered in systemic terms, taking account of the potential for optimal co-delivery of a range of linked ecosystem service benefits. The second then is to plan for and demonstrate the multiple values of optimisation of overall societal benefits achieved through systemic solutions. 'Systemic solutions' thus conceived are defined as "…*low-input technologies using natural processes to optimise benefits across the spectrum of ecosystem services and their beneficiaries*".[7] Wetland, washland and urban ecosystem-based technologies, optimised to achieve multiple benefits simultaneously, are explicitly recognised under this initial definition. However, many more multi-beneficial solutions are possible, including those promoted by the UK's Natural Capital Committee but also though reforestation, catchment management for water resource conservation, simultaneously with water quality control or flood regulation, 'green infrastructure', strategic coastal and marine management, ecosystem-based farming, or resource stewardship as a sustainable basis for business. These twinned concepts—anchor services and systemic solutions—encourage optimisation of net societal value across all ecosystem services, without skewing benefits towards a favoured few at cost to other generally overlooked beneficiaries, particularly including future generations. Net societal value thereby is secured, including the resilience and continued functioning of the benefit-providing ecosystems.

Maybe this all sounds like an unattainable, utopian pipe-dream, impossible to achieve given the fixed workings of markets and societal rights and the increasing pressures of the growing human population and climate change? But it is the very fixedness of the norms we have instigated that drive the degenerative socio-ecological cycles that we see everywhere across the world today. It is clear that those investing in short-term asset exploitation tend to benefit disproportionately, with the resource-liquidating habits deeply enshrined in a neoliberal economic model shaping policy and business practices across increasing areas of the globe. It is equally abundantly transparent that the dispossessed suffer asymmetrically under a destructive model, in which all will inevitably ultimately

[7] Everard, M. and McInnes, R.J. (2013). Systemic solutions for multi-benefit water and environmental management. *The Science of the Total Environment*, 461(62), pp. 170–179.

succumb to resource poverty and potential conflict. Is this a vision we actually want, or of which we can feel proud?

The alternative vision of a more connected world view needs progressively to supersede the patchworks of narrowly-framed technical, legal and fiscal 'fixes' that are currently often badged as sustainable development. Of course, any form of societal transformation, including that founded on achievement of sustainability, challenges vested interests and fixed assumptions. One particularly stark reality that this confronts is that property rights tend to be physical and static, benefitting private individuals, whereas ecosystem functions and services tend to provide public benefits and to flow across broad landscapes and longer-term futures horizons, regardless of human-drawn deeds and lines on maps. Add to this the inordinately greater complexity of a contemporary world view in which multinational corporate entities, with more wealth than whole smaller nations, extend global fingers of resource exploitation and political influence to all corners of the world. The legacy policy environment shaping relationships along these value chains is as yet barely influenced by systemic thinking and the optimisation of outcomes.[8] Much of it remains firmly rooted in industrial-era assumptions about the inexhaustibility of resources, externalising the wider costs of waste, resource depletion and implications for non-consumers.[9]

It is this fragmented environment of policies, economics and vested interests that needs to be shaped by lessons from the preceding overview of 'regenerative landscape' case studies. Fiscal and other major legislation reforms are required to bridge this current divide by constructing reward systems, backed up where necessary by compulsions, with net societal security and wellbeing as priority outcomes. Agri-environment subsidies already go some of the way, albeit currently in a rather imperfect and marginal way, by directing a small proportion of public funds into management of farmland aimed at delivering desirable societal outcomes. The same is true of the slow transition towards a landscape-based approach to Natural Flood Management, and the designation of 'no take zones' and other marine protected areas enabling natural regeneration of marine biodiversity including fish stocks. At a global scale, REDD+ is an example of a scheme direct-

---

[8] UK National Ecosystem Assessment. (2011). *The UK National Ecosystem Assessment: Synthesis of the Key Findings*. UNEP-WCMC, Cambridge.

[9] Jackson, T. (2011). *Prosperity Without Growth: Economics for a Finite Planet*. Earthscan, Abingdon.

ing payments from already-developed nations into conservation and regeneration of forest in developing nations, in large measure to mitigate climate change but also protecting indigenous livelihoods and rights.

We can and must rise to this challenge of societal transformation. The evidence base presented in the preceding section of this book shows how we are doing so already in scattered microcosms across the world, in a diversity of habitat types, and in nations at all stages of development. We need now the vision and concerted commitment to reach a 'tipping point' into new social norms across society, ultimately beneficial to all.

4.2 Tools for Systemic Understanding and Action

It is generally not lack of good intention that degrades ecosystems vital for tomorrow's wellbeing. Rather, it is lack of integrated oversight across the whole connected socio-ecological system, including the policy and economic forces that shape it. Just one of the examples reviewed previously is that of opening up tillage on prairie lands as an economic stimulus to avert the worst outcomes of the Great Depression, a laudable aim yet one that inadvertently culminated in the disastrous and ruinous American 'Dust Bowl'. This is an extreme example amongst many about how good intentions, such as market-driven production of cheap food, can damage vital soil, water and biological resources undermining human security and opportunity.

The STEEP framework (an acronym for its five constituent social, technological, economic, environmental and political elements) has previously proven useful for addressing linked environmental and social implications in technological, political and economic contexts. The STEEP model was initially developed to encourage broader thinking in business strategy beyond established assumptions, ensuring multiple external factors impacting an organisation are considered in addressing global change.[10] However, it has since proven useful for exploring systemic relationships in other domains of human activity,[11] including addressing sustainability

---

[10] Morrison, J. and Wilson, I. (1996). The strategic management response to the challenge of global change. In: Didsbury, H. (Ed.), *Future Vision, Ideas, Insights, and Strategies*. The World Future Society, Bethesda, MD.

[11] There are variations on this model, such as PEST, PESTEL, PESTLE, STEPJE, STEP, STEEPLED and LEPEST.

goals.[12] Whilst STEEP can be applied as a simple classification scheme, it can also be used as a systems model in which interdependencies between all five constituents are considered, setting social and environmental interdependencies in complex economic, technological and political contexts. STEEP has proven invaluable in understanding the wider implications of deployment of technology and governance systems in management of water, ecosystem service flows and dependent development issues in South Africa,[13] Europe,[14] India[15] and elsewhere.[16]

Implicit in systemic appreciation, in this case as articulated by application of STEEP as a systems model, is that all dimensions of the system are integrally connected. For example, there is no such thing as a purely 'social problem' as every social context is supported by and has implications for ecosystems and their services, for which the distribution of benefits and potential costs have economic implications, all shaped by the formal and informal governance (policy) context in which the issue arises and any technological solution devised to address it. The same observation applies to 'environmental problems' that are as often technology-, economic- or policy-driven with distributional impacts across society now and into the future. Synergistic and antagonistic outcomes can arise within any STEEP category in the absence of a systemic framework for thinking and implementation, as for example in widespread deployment of unregulated energised tube wells depressing groundwater levels, rendering other less intensive techniques, such as open wells and traditional water-lifting techniques, unable to access receding water resources. All elements of the STEEP framework, applied as an interdependent systems model, follow this same pattern of interconnectedness, as illustrated by the examples in Table 4.1.

---

[12] Steward, W.C. and Kuska, S. (2011). *Sustainometrics: Measuring Sustainability, Design, Planning and Public Administration for Sustainable Living.* Greenway Communications.

[13] Everard, M. (2013). *The Hydropolitics of Dams: Engineering or Ecosystems?* Zed Books, London.

[14] Everard, M., Harrington, R. and McInnes, R.J. (2012). Facilitating implementation of landscape-scale integrated water management: The integrated constructed wetland concept. *Ecosystem Services*, 2, pp. 27–37.

[15] Everard, M. (2015). Community-based groundwater and ecosystem restoration in semi-arid north Rajasthan (1): Socio-economic progress and lessons for groundwater-dependent areas. *Ecosystem Services*, 16, pp. 125–135.

[16] Everard, M. and Quinn, N.W. (forthcoming). Landscapes that hold water: Contrasting perspectives and opportunities from perceived dry (Rajasthan) and wet (England) climatic zones.

	Social	Technological	Environmental
Social	• Asymmetric power can result in inequitable outcomes by marginalisation of the many by a privileged few • Conversely, an inclusive approach can create greater distribution of benefits across society		
Technological	• All technological solutions have socially distributional outcomes that need to be recognised and incorporated in technology choice and management • Any remaining externalities for people arising from technology use need to be considered and mitigated • Participation of all potentially affected stakeholders in decision-making helps recognise, find solutions to and gain acceptance for technical solutions	• Intensively exploitative technologies (such as tube wells and dams) can make other technical approaches (such as open wells) ineffective • Conversely, hybridisation of technologies to address inevitable externalities can be tuned to the supportive processes of ecosystems (such as emerging thinking about hybridised water exploitation in the Banas catchment of Rajasthan)	

Table 4.1 Interrelated aspects socio-ecological challenges using the STEEP framework

				Economic	Political
Environmental	• The distribution of benefits and disbenefits across the full spectrum of ecosystem services, including internal ecosystem resilience, is essential for attaining equity as well as ecosystem resilience	• Consequences across the full spectrum of ecosystem services, including internal ecosystem resilience, is an essential consideration for sustainable technology deployment	• Narrow exploitation to maximise a favoured ecosystem service can degrade ecosystem functions and linked ecosystem services, as frequently seen in contemporary intensive agriculture and capture fisheries • Conversely, if the focus is on underpinning ecosystem processes, 'systemic solutions' can protect and optimise all ecosystem services		
Economic	• Distribution of both benefits and disbenefits across different stakeholder groups, as well as of scheme costs, is essential to account for full societal value	• Distribution of both benefits and disbenefits, now and into the longer term, is an important consideration to inform sustainable technology choice • Any remaining externalities arising from technology choice need to be considered and mitigated to optimise net societal value	• Ecosystem services need to be considered on a fully systemic basis to achieve optimal net societal value • The fiscal system may need revision to accommodate a systemic perspective	• Where benefit assessment looks only at narrow outcomes, or perverse taxes or subsidies result in unintended impacts on ecosystems and people, overall distributional inequities in benefits and costs are likely to arise • Conversely, an inclusive approach can create greater distribution of benefits across ecosystem services and society (as in poverty alleviation strategies in China's Loess Plateau and the Ethiopian Highlands)	
Political	• A clear recognition of who wins and who loses from policy decisions is essential to address systemic, more sustainable outcomes • Participation of all potentially affected stakeholders in decision-making helps recognise, find solutions to and gain acceptance for governance decisions	• The formal and informal policy environment has a major impact on technology choice, so may need reform if options appraisal of technical solutions are to become sustainable	• Implications of policies for the full range of ecosystem services is vital not only for societal equity but also to avoid inadvertently degrading supporting ecosystems • Policy reform to promote ecosystem restoration as a primary natural capital is generally economically beneficial, or alternatively may become a source of potential conflict	• Implications of policies for the full range of ecosystem services and their associated benefits and beneficiaries (or potential costs and victims) need to be accounted for in policy to guide sustainable decision-making	• Decision-making frameworks shaped by privileged sectors of society tend to marginalise many in society, including the resilience of supporting ecosystems upon which all depend • Conversely, a participatory approach can reflect all interests and forms of knowledge to generate more robust and better-accepted decisions

Table 4.1   (continued)

4.3   Connections and Disconnections

Beneath all of the examples of 'degenerative landscapes' addressed in this book is a systemic failure either to account for one or, generally, more of these STEEP constituents, or to overlook interdependencies between them (see examples in Box 4.1). Conversely, underpinning all of the examples of 'regenerative landscapes' is a cohesive approach to consideration of all these factors (see Box 4.2).

Box 4.1 An Indian Water Management Example of 'Degenerative Landscapes', Assessed by STEEP Constituents

Systemic failures of the India's recent technocentric water policy environment driving degenerative cycles, outlined in the Introduction of this book, stem from:

- Social factors include the shift from communal water management to a private benefit, competitive model related to mechanisation of water extractions;
- Technological factors such as proliferation of unregulated tube wells as well as large dam-and-transfer schemes focus on technically efficient extraction without rebalancing stewardship of the natural resources;
- Environmental factors include the degradation of ecosystem resilience as water systems are over-abstracted, also overlooking the multiple flows of diverse ecosystem services stemming from well-functioning water systems;
- Economic factors include a narrow framing of short-term profit against long-term societal value, overlooking distributional impacts across all ecosystem services and their beneficiaries, exacerbated by subsidised energy promoting wastage of water as well as health issues related to consumption of increasingly geologically contaminated water from deeper groundwater; and
- Political factors include formal and informal governance systems that are narrowly focused on economic growth, and which favour intensive water users over and above sustenance of supporting ecosystems and rural constituencies more closely dependent upon them.

> **Box 4.2 An Indian Water Management Example of 'Regenerative Landscapes' Assessed by STEEP Constituents**
>
> Regeneration of water, soil and livelihoods in Alwar District, Rajasthan, were described as:
>
> - Social factors include restored community institutions and collaborative practices accounting for all villagers' needs;
> - Technological factors such as restored or innovated johadi, anicuts and check dams that work with natural processes;
> - Environmental factors focus on regeneration of water systems and soil fertility as the primary resources for societal wellbeing;
> - Economic factors address security and opportunity for all, rather than favouring a narrow subset of influential players; and
> - Political factors include governance arrangements founded on community agreements on water management and benefit-sharing, both at village and catchment scales.

4.4   Leadership for Sustainable Change

There is broad acknowledgement about the increasing urgency of precipitating wholesale transformation of developed world society, and the influence it exerts on developing regions. At the heart of this necessary transformation is restoration of the supportive ecosystems underpinning continuing human wellbeing as the primary focus of deployment of technologies, economic measures and policy reform. Innovation, appraisal and implementation of appropriate and sustainable technologies may then be determined by both geographical and cultural factors, in which people are acknowledged as key resource owners, actors and stewards. This then requires tiered governance arrangements, supported by economic incentives and sanctions to optimise net societal value when all ecosystem services are considered. But how is this systemically connected societal transformation driven?

There may be an automatic assumption that government leads societal change through the setting of legislation. However, this is not necessarily, and in fact may rarely be, how change actually happens. Leadership for sustainable change can and does happen from amongst all actors across the social system. One lens through which to view the influence of these sectors in fomenting change is by examination of the four principal subdivisions of social systems: public (government); private (business); academic; and civil society sectors. Table 4.2 gives examples of leadership

Table 4.2 Examples of systemic leadership from different societal sectors

Societal sector	Example of leadership for systemic change
Public (government)	• Local government is a key player at a local scale in the Stroud Rural SuDS Project. • National government and its agencies have undertaken a leadership role in promotion of a Natural Flood Management approach. • Government and its regulators, together with NGOs, water companies and knowledge support from academia, have played leading roles in a landscape-based approach to raw water resource management, as for example New York City's landscape-based protection of its unfiltered water supply system. • Once accepted and supported by other players in society, government has a major role to play in cementing new societal norms through legislation, fiscal systems and strategies. We will return to governance of sustainability later in this chapter, considering its pivotal roles in greater detail and dimension.
Private (business)	Many regenerative examples of corporate leadership have been expounded in the preceding chapter of this book, some illustrative examples of which include: • Industries dependent on primary resources—agriculture, forestry, food and drink and their regulators—are amongst the pioneers in development of stewardship schemes such as FSC, MSC, Rainforest Alliance and other approaches seeking to create new markets based on resource protection and renewability. • Water companies have been leading partners, with NGOs and state regulators, in a landscape-based approach to raw water resource management, as for example the Upstream Thinking and SCaMP programmes. • Other highly water-dependent businesses can also lead, highlighted by approaches advertised by Diageo, SABMiller, PepsiCo India and Coca Cola to promote water recharge and security. • Ecosystem-centred farming methods at Loddington Farm, Hope Farm, Knepp Estate and the Kellogg Biological Station demonstrate simultaneously profitable and regenerative farming methods using off-the-shelf technologies. • Some business consortia also work together as innovators for sustainability such as, in the UK, the Aldersgate Group[a] and Business in the Community (BiTC)[b] and, globally, the World Business Council for Sustainable Development (WBCSD).[c]

(continued)

Table 4.2 (continued)

Societal sector	Example of leadership for systemic change
Academic	• Academia has a key role to play in innovation, particularly in relation to sustainability challenges for which the private sector may not yet be ready to invest.
	• Academics also have key roles to play in identifying, testing and providing robust evidence to promote sustainable solutions, for subsequent transfer to policy and practice.
	• Though not perhaps narrowly 'academic, 'Think Tanks' of various types also play important thought-leading roles, including both knowledge generation and transfer into practice. The Club of Rome[d] is one such international body of thought leaders—notable scientists, economists, businessmen and businesswomen, high level civil servants and former heads of state from around the world—who share a common concern for the future of humanity, striving to make a difference by promoting understanding of the global challenges facing humanity and their potential solutions through scientific analysis, communication and advocacy.
Civil society	• NGOs are often key actors in harnessing diverse interests across social groups, merging them with traditional and other forms of knowledge to challenge established norms including perverse policies, promote effective solutions and spread learning.
	• Ultimately, social engagement is a key determinant of successful transition as expressed by former US President Barack Obama: *"Change happens when ordinary people get involved"*.
	• People, acting via whatever sector of society, are the key change agents; every one of us has more power and influence than we might commonly assume.
	More consideration and examples of 'people power' follow below.

[a]http://www.aldersgategroup.org.uk/
[b]https://www.bitc.org.uk/
[c]https://www.wbcsd.org/
[d]https://www.clubofrome.org/

from each of these sectors, though noting that the 'mainstreaming' of issues tends to occur when their implications span multiple societal sectors.

4.5   People Power

Ultimately, people are principal actors, as well as beneficiaries or victims, of the use or abuse of supportive ecosystems. The purpose of government, in theory, at least, is to serve the wellbeing of the people it represents and who, let us not forget, pay the salaries of those transacting governance. The purpose of the economy is as a means of exchange between recipients and providers of goods and services, mostly sourced as primary material and energetic resources from nature as well as having impacts on natural systems that in turn affect the wellbeing of citizens. Academia serves as a knowledge-generating and disseminating hub, supportive of economic and policy progress in support of human wellbeing. It is the grand intention that all of humanity is central to the deliberations of politics, the market and thought leaders, though this is hard to discern in the everyday practice and apparent self-serving nature of contemporary markets and politics. People are certainly central to sustained change—quoting former US President Barack Obama again, "*Change happens when ordinary people get involved*"—as instigators, protagonists and in the acceptance of change.

Populist instigation of what became known as the 'environment movement' can be traced back into pre-history in traditional beliefs, protocols and taboos, rituals and obligations. In the modern world, environmental awareness can be traced back to civil actors such as in the late nineteenth century, for example with publication of Henry David Thoreau's *Walden; or, Life in the Woods* in 1854[17] in the United States. In Britain, concern about the hunting of great-crested grebes to near extinction (reduced to just fifty breeding pairs by 1860) to serve ladies' fashions mobilised public and political opinion leading to foundation of the Society for the Protection of Birds between 1889 and 1891, and its subsequent granting of a Royal Charter to become the Royal Society for the Protection of Birds (RSPB) in 1904. In 1905, the National Audubon Society was incorporated in the

---

[17] Thoreau, H.D. (1854). *Walden; or, Life in the Woods*. Ticknor and Fields, Boston.

USA, a civil institution named in honour of the French-American orni-thologist John James Audubon for his prophetic warnings about the threats of over-hunting and habitat loss for many once-common American birds.

A particularly intense period of civil movements in and around the 1970s brought 'the environment' from marginal concern into mainstream political discourse. This was a time of many public 'grassroots' protests, learned publications such as the Club of Rome's seminal *The Limits to Growth*[18] report in 1972, and the international *United Nations Conference on the Human Environment* held in Stockholm, Sweden, from 5th to 16th June also in 1972. From personal experience from my involvement, I can vouch that a sense of outrage against a stolen future, as well as the moral offence of our treatment of the planet's ecosystems, drove much of the environmental protests of the 1970s, paving the way for formerly margin-alised environmental issues to progressively enter the political mainstream.

These are just some of the many events and publications in the longer-term transformation of industrialised society from one ignorant of its environmental obligations and dependences towards one of emerging awareness and hesitant transition to safeguard the future for both inher-ent and human benefits. Influential public activism continues today, motivated by individual or collective outrage about use, abuse and ineq-uity in the sharing of nature, supportive natural resources, climate change implications and future human opportunity. Box 4.3 outlines just a few civil movements of high public profile in 2019, as I write this book, all pertaining to aspects of how our treatment of the atmosphere and other critical natural resources is compromising future life prospects.

Box 4.3 Some Examples of 'People Power' in 2019

Reaction to societal inaction, likely to condemn the rising generation to a dystopian future of increasing threats and limited life opportunities, under-pin many emergent civil movements in 2019, including:

- Extinction Rebellion is an international movement using non-violent civil disobedience in an attempt to halt mass extinction and minimise the risk

---

[18] Meadows, D.H., Meadows, D.L., Randers, J. and Behrens, W.W. (1972). *The Limits to Growth.* Potomac Associates, Universe Books.

of social collapse.[19] Its foundation was around a public protest in Parliament Square in London during October 2018, for which an unexpectedly large crowd gathered. This led on to more protects, with increasing numbers of participants. Extinction Rebellion groups have since sprung up in dozens of countries from the Solomon Islands to Australia, from Spain to South Africa, the US to India, using non-violent protests to agitate for disruptive political change.

- Swedish teenager Greta Thunberg became globally infamous during 2019 for her outspoken demand—initially protesting alone outside the Swedish parliament in 2018 but now with global reach—for immediate action to combat climate change. She has since become an outspoken climate activist with an international profile that has seen her initiating a 'school strike for climate' movement joined by an estimated 1.4 million students in 112 countries as of March 2019.[20] In March 2019, three deputies of the Norwegian parliament nominated Thunberg for the Nobel Peace Prize[21] and, in May 2019 at the age of 16, she featured on the cover of *Time* magazine.[22]

- 'Generation Z' (or 'Gen Z') is a marketing term describing a demographic cohort typically born between the mid-1990s to mid-2000s. These young people, users of the Internet from a young age, are comfortable with technology and social media, and are reaching voting age. Amongst this generation, an emerging individual and collective movement using the name 'Generation Z' is participating in increasing activism to fight for a better future, including on issues such as gun violence, the opioid epidemic, transgender rights and climate change. This generation has been brought up with the reality of a future less advantageous than that enjoyed by previous generations, living through turbulent world events such as the great recession, potentially debilitating climate change predictions, mass shootings, biodiversity decline and low confidence in conventional media due to a bombardment of 'fake news'. The term 'Generation Z' has also been redefined as representing the last generation that can make a positive change in society as its natural resource base degrades increasingly rapidly under legacy industrial world habitats, from which this generation will increasingly become victims rather than beneficiaries.

---

[19] https://rebellion.earth/the-truth/about-us/.

[20] Shabeer, M. (2019). Over 1 million students across the world join Global Climate Strike. *Peoples Dispatch*, 16 March 2019.

[21] Vaglanos, A. (2019). 16-year-old climate activist Greta Thunberg nominated for Nobel peace prize. *Huffington Post*, 14 March 2019.

[22] Plant Based News. (2019). Greta Thunberg on the cover of TIME: "Now i am speaking to the whole world". *Plant Based News*, 16 May 2019.

The key factor in these considerations of 'people power' is that all describe how people have acted, and can act whether individually or together, to motivate change. People—all of us—possess more power and influence than we might commonly assume, or that we might wield in shifting opinion and practice for the greater benefit of all.

4.6   Governance for Sustainability

Having noted that leadership can come from all sectors of society, it is important also to recognise that governance systems of multiple types have great significance as enablers or blockers for progress with sustainable development. One of the strengths of the STEEP approach is that it embeds socio-ecological interactions within the complex system of political and economic forces, including those pertaining to technology choice. Addressing issues and decisions on a fully systemic basis within the complexity of cultural systems is fundamental for sustainability and regeneration, for which the 'political' strand of the STEEP model is of great importance as it is ultimately the decisions we make and the choices and actions that follow them that determine regenerative or degenerative outcomes across the whole system.

Political considerations include not just formal, binding rules and regulations, but also diverse governance arrangements spanning informal and often traditional and cultural protocols, in addition to consensual agreements from local to intergovernmental scales. There is connectivity and cross-fertilisation across these scales, for example in the manner in which the Government of Rajasthan's *Mukhyamantri Jal Swavlamban Abhiyan* (MJSA) programme promoting community-scale water harvesting and governance has now built on millennia of locally-based traditional practices and the more recent efforts of NGOs to restore them. Equally, high-level, international agreements on tackling climate change lack substance without resolution into national policies and practical local actions.

In densely populated and developed world situations, a backbone of statutory regulations plays crucial roles, though these again may be positive or negative. Positively, evolution of legislation to enforce environmentally and socially responsible practices is essential to embed new understanding and priorities into cross-societal norms, as also to prevent less scrupulous companies from gaining market advantages by undercutting competitors. The same principle applies multi-nationally, emphasising the need for international rules on trade, and ratification of pro-environmental and pro-social conventions. However, vested interests benefitting from less enlightened legacy legislation impose 'regulatory drag', slowing mainstream integration of rules governing more sympathetic practices to safeguard, or ideally to restore, supporting ecosystems. Consequently, visionary leadership based on evidence-based and transparently stated longer-term goals is essential if we are to aspire to a regenerative and sustainable future.

The frequent lack of such vision-led leadership calls into question the sincerity of governments in acting in the best interests of global society, significantly including those of future generations, as opposed to protecting the privileged few within and between nations. That the dominant global political model is broken in this regard is hardly a novel sentiment, but setting this in the context of our relationship with ecosystems vital for continuing human wellbeing amplifies its significance. In many ways, 2015 constituted a high water mark of multilateral global intent around the fundamental relationship between humanity and the ecosystems essential to sustain it. This took the shape of three high-profile intergovernmental agreements: the Sendai Framework on Disaster Risk Reduction; the UNFCCC Paris Agreements; and, perhaps most significantly, the UN Sustainable Development Goals. However, wind forward just one year, and the UK's Brexit referendum and President Trump's election win in the US seemed to trigger a global cascade of unilateralism, neoliberal economic protectionism, retreat from consensual international conventions and other forms of fundamentalism, puncturing the fragile bubble of optimism. How could this reversal have been permitted by an increasingly environmentally literate electorate, and when the key role of civil government is, as David Hume wrote in 1739, to overcome individual *"…narrowness of soul, which makes them prefer the present to the remote"*?[23] Regrettably, the short-termism of the electoral cycle creates equally short political time horizons, reinforced by business and other special interest

---

[23] Hume, D. (1739). *A Treatise of Human Nature*. John Noon, London.

lobbying founded on maintaining a status quo, leading to short-term gains and pre-election 'sweeteners' as political parties seek primarily to retain power rather than to secure society's longer-term best interests through driving transitions that can work to arrest and reverse ecosystem degradation. Most fundamentally, this democratic model completely overlooks the rights of future generations, with global society in the post-territorial colonial era now effectively colonising time instead by creating future resource limitations, 'dumping' our long-term pollutants there and seriously reducing ecosystem integrity and capacities vital for sustaining future generations.

The dysfunctionality of today's dominant political model in this regard is increasingly evident. Yet, like the examples of regenerative uses of land and waterscapes, there are also green shoots of potentially regenerative political practices, without appealing to the flawed model of a 'benign dictator'. (The global history of dictatorship teaches us that initially bright visions tend rapidly and seemingly inevitably to collapse into nepotism of various kinds.) Box 4.4 outlines a diversity of models from across the world in which political systems have introduced some steerage, of varying efficacy, to take better account of a future in which the needs of people and the supportive capacities of ecosystems are in reality indissoluble.

Box 4.4 Models Offering Longer-term Steerage of Political Decisions

Wales established a Future Generations Commissioner under the Well-being for Future Generations Act of 2015, the Commissioner's role being to ensure that public bodies make policy decisions looking at least 30 years into the future.

There are calls for a similar Future Generations Act to cover the entire UK. Some early progress is already being made in the form of the establishment in 2018 of an All-Party Parliamentary Group for Future Generations.

Finland implemented a parliamentary Committee for the Future in 2017, tasked with scrutinising legislation for its impact on future generations.

Israel created the role of Ombudsman for Future Generations in 2001, though regrettably the position was abolished in 2006 as it was perceived as too influential in delaying legislation. This does not bode well for the 'mainstreaming' of future concerns in Israeli politics.

A municipality-level movement in Japan known as Future Design comprises citizen assemblies in which participants play roles respectively representing current and future citizens. Ultimately, the movement aims to establish a Ministry of the Future in central government and Departments of the Future in local government authorities, with policy formulation informed by the future citizens' assembly model.

In the US, a youth-led organisation, Our Children's Trust, is attempting to secure the legal right to a stable climate and healthy atmosphere for the benefit of all present and future generations.

One particularly high-profile outcome in Wales resulting from intervention by the Future Generations Commissioner, established under the Well-being for Future Generations Act of 2015, was the scrapping in early 2019 of a £1.4 billion relief road scheme for the M4 motorway that had been planned to cut across the Gwent Levels in south east Wales (Box 4.5). However, this example also demonstrates that robustly argued pro-environmental decisions safeguarding ecosystems of inherent value and underpinning longer-term human wellbeing do not come without opposition from narrower, short-term business interests. This was the experience in Israel, leading to the abolition of its Ombudsman for Future Generations in 2006 due to perceived delays to the passage of legislation created by consideration of its long-term consequences.

Box 4.5 Scrapping of the Planned M4 Relief Road Through the Gwent Levels, South Wales

A high-profile decision in Wales in 2019 with significant intervention of the Future Generations Commissioner, under the Well-being for Future Generations Act of 2015, was the scrapping of a £1.4 billion relief road scheme for the M4 motorway that had been planned to cut across the Gwent Levels in south east Wales.[24]

The ecological and historic significance of the Gwent Levels, home to wetland species such as otters, water voles, scarce insects, plants and wading birds, and with drainage reens (ditches) built by the Romans, is reflected in designation of a cluster of six SSSIs (Sites of Special Scientific Interest). These SSSIs collectively span 5,700 hectares (14,000 acres) alongside the Severn Estuary, covering the whole of the Gwent Levels between the east of Cardiff through to Caldicot.

The decision to abandon the planned motorway scheme was criticised by business leaders and the Conservative Party for blocking moves to address

---

[24] Morris, S. (2019). Wales scraps £1.4bn Gwent Levels M4 relief road scheme. *The Guardian*, 4 June 2019. https://www.theguardian.com/uk-news/2019/jun/04/wales-scraps-gwent-levels-m4-relief-road-scheme.

congestion on a principal route into South Wales, affecting Welsh commut-
ers and business people. However, the final determination, following a
public enquiry, was based not only on cost grounds but also consideration
of the seriously deleterious effect of carving a 14-mile, six-lane stretch of
motorway around the city of Newport, wrecking a precious ecological and
historic landscape.

Ultimately, a key issue is to take such decision-making, or at least the
checks-and-balances upon decision-making, out of the realm of often-
divisive, short-term party politics. This is particularly so for two-party political systems, in which the ideology of one party tends to become
automatically opposed by that of the other. This is well-illustrated by the
actions of the Trump Republican Party presidency in the US from 2017,
dismantling much of the progressive social legislation (particularly relating
to access to health care) and commitments to international environmental
protocols implemented under the preceding two terms of the Obama
Democratic Party presidency. Some short-term reactive political responses
may be beneficial, such as global mobilisation around marine plastic accumulation in the latter half of the 2010s leading on to a more fundamental
rethink of resource use and waste strategies. However, this is very much an
isolated example of good outcomes against a wider terrain of short-term
reactivity blind to long-term outcomes. A stark example of short-term reaction with long-term costs is the UK's abandonment of renewable generation subsidies (Feed-in Tariffs) in the name of short-term austerity,
constricting progress towards longer-term carbon neutrality with all its
breadth of potential health, social and environmental benefits, averted
costs, and innovation leading to lucrative patent ownership. Reform of generally tightly linked political and economic components is of fundamental
importance for achievement of longer-term social and ecological security
and opportunity. The 30-year views required under the Future Generations
Act 2015 in Wales is a pragmatic starting point for this, extending the horizon beyond short-term electoral cycles or single generations, and taking
better if not perfect account of the slower pace of many ecosystem responses.

A further example of taking decisions out of the hands of partisan politics
comes from the UK, where the government gave the Bank of England
greater independence in 1997 particularly relating to the setting of interest
rates, meaning that modification of this rate could no longer be used directly

as a political tool. In 2010, the UK government went further by establishing an Office for Budget Responsibility as a non-departmental public body tasked with providing independent economic forecasts and analysis of the public finances, cemented on a statutory basis by the Budget Responsibility and National Audit Act 2011. Though there is contention around these transitions in the handling of public finances relating to detachment of decisions from the accountability of politicians, they do at least make progress with the de-politicisation of decisions and views with longer-term implications. Also in the UK, the Water Industry Act 1991 requires each private water company to produce a Water Resources Management Plan (WRMP), updated every five years but addressing a minimum 25-year planning period, with the aim of ensuring that there is sufficient supply of water to meet the anticipated demands even under dry conditions when supplies are stretched and demand for water tends to be higher than normal. Taking more of these types of decisions and reviews relating to foundational socio-ecological systems out of the contested ground of short-term, adversarial politics, and accounting for holistic outcomes rather than narrow government Departmental priorities or private sector return-on-investment cycles in isolation, is a fundamental requirement if society is to make transformational progress to secure a sustainable or ideally regenerative future.

We can also learn from the outcomes of tribal and other forms of governance in which people make decisions informed by understanding of their relationships with the ecosystems they inhabit, and in which they have collective ownership. The seriously negative trends in biodiversity reported globally in the May 2019 Global Assessment Report[25] of the Intergovernmental Science-Policy Platform on Biodiversity and Ecosystem Services (IPBES) were observed to be, on average, less severe or avoided in areas held or managed by Indigenous Peoples and Local Communities. As at least a quarter of global land area is traditionally owned, managed, used or occupied by Indigenous Peoples, this observation is significant, though deterioration under local livelihoods was nonetheless observed across much of this area albeit at a slower pace. Examples from two contrasting regions of the Indian Himalayas under different forms of governance, yielding significantly different outcomes for livelihood-ecosystem interdependencies,

---

[25] Brondizio, E.S., Settele, J., Díaz, S. and Ngo, H.T. (2019). *Global Assessment on Biodiversity and Ecosystem Services of the Intergovernmental Science-Policy Platform on Biodiversity and Ecosystem Services (IPBES)*. IPBES. https://www.ipbes.net/global-assessment-biodiversity-ecosystem-services.

amplify the potentially significant role played by community ownership and governance (Box 4.6). The knowledge, innovations and practices, institutions and values of indigenous peoples and local communities, particularly their connectedness with the landscapes that supported their history and are recognised as foundational to their future wellbeing, contain lessons for the global community about wise environmental governance (Fig. 4.1).

Box 4.6 Contrasting Governance and Outcomes for Livelihood-Ecosystem Interdependencies in the Indian Himalayas

Study of mountain villages in the Indian Himalayan state of Uttarakhand revealed low food availability and decreasing self-sufficiency, under the combined pressures of increasing foraging by wildlife coupled with seasonal to permanent outmigration by younger men seeking more secure income and alternative livelihoods.[26] Whilst residents respected the wildlife compromising their livelihoods, statutory prohibitions on disturbing wild animals and exploiting forest resources limited livelihood options. Furthermore, much of the income remitted by migrants to their villages was not retained locally, but flowed back out of the Himalayan region through purchases of food produced and marketed in the plains, threatening the economic viability of villages whilst placing asymmetric pressures on remaining female, elderly and young people who concentrate labour on local livestock production to the neglect of crop agriculture.

Tribal rights are protected in the north-eastern Indian state of Arunachal Pradesh, conferring a high degree of autonomy through village-scale governance arrangements that substantially limit the influence of state and private actors on natural resource exploitation. A case study in Lileng village revealed a close, synergistic understanding and relationship between local people and the natural ecosystems for they have protected rights. They make diverse uses, enjoy a high dependency and manage them on a sustainable basis that retains dense forest and unconstrained river flows.[27] Tribal priorities oppose state- and national-level aspirations to modify the river system by damming, and resist the intrusion of industry, focusing instead on the community's subsistence needs. There was a high level of food sufficiency and natural resource protection, with low out-migration and gender/age inequalities, attributed to tribal rights and authoritative local governance.

---

[26] Everard, M., Gupta, N., Scott, C.A., Tiwari, P.C., Joshi, B., Kataria, G. and Kumar, S. (2018). Assessing livelihood-ecosystem interdependencies and natural resource governance in Indian villages in the Middle Himalayas. *Regional Environmental Change*, 19(1), pp. 165–177. https://doi.org/10.1007/s10113-018-1391-x.

[27] Everard, M., Kataria, G., Kumar, S. and Gupta, N. (forthcoming). Assessing livelihood-ecosystem interdependencies and natural resource governance in a tribally controlled region of India's north-eastern Middle Himalayas.

Fig. 4.1 The largely intact mountain rain forest surrounding Lileng village and the Semang River that is fed from it is under the stewardship of the Adi tribe, the north-eastern Indian state of Arunachal Pradesh enjoying protected tribal rights and autonomous village governance systems which maintain healthy ecosystems valued for the many ways they support livelihoods. (Image © Dr Mark Everard)

4.7   Evolution or Revolution?

Change at a grand, cross-societal and global scale is required. But is this revolution or evolution? This is an idea I grappled with in my 2016 book *The Ecosystems Revolution*,[28] concluding that violent overthrowing of regimes produces anarchy rather than the kind of harmonised transition we now require. However, a strategic, or sometimes serendipitous, series of evolutionary changes tend to be what we look back on as a period of 'revolution'. Let us not forget that the Industrial Revolution in Europe describes a long-term transition to new manufacturing processes from about 1760 to around 1840. Likewise, the European Agricultural Revolution spanned over three centuries of incremental innovation rather than swift or orchestrated change. Industrial and agricultural innovations continue today, with some countries just embarking on their own 'revolutions'. Neither of these 'revolutions' was choreographed, but the kind of revolution we now urgently need to reconnect us with the ecological foundations of a sustainable future will certainly need strategic guidance.

We are undoubtedly seeing major changes across society in response to growing environmental literacy, including the many examples of 'regenerative landscapes' reviewed in the preceding chapter. However, perhaps the biggest constraint upon transformative change is the tight grip of neoliberal market economics, consumerism constituting one of the most globally pervasive ideologies throughout human history. Contemporary business models and markets are essentially a product of the European Industrial Revolution, innovated as a means of exchange upon discovery and implementation of innovative technological means for the large-scale transformation of basic resources into useful products. But markets are inherently amoral and, unconstrained by policies and fiscal measures to buffer their impacts on supportive ecosystems and inequities in the sharing of benefits and costs across society, tend to turn into a beast that eats itself by maximising immediate profit at minimal cost. This inherently

---

[28] Everard, M. (2016). *The Ecosystems Revolution: Co-creating a Symbiotic Future*. Palgrave PIVOT Series. Springer, Switzerland.

suicidal tendency is at the heart of the neoliberal approach to economics embedded and spread increasingly globally during the 1980s under the leadership of the Thatcher government in the UK and the Reagan presidency in the US. Today, we seek change within the constraints of a market system that inherently marginalises ecosystems, the renewability of natural resources, and equality of rights across society including in particular the wellbeing and needs of future generations.

Various initiatives are seeking to challenge the supremacy of markets, and in particular the exclusion of nature and its supportive ecosystems. Some of these approaches include UNEP's 'New Deal for Nature' (Box 4.7), the Regenerative Communities Network (Box 4.8) and, at national scale, the Regenerate Costa Rica Hub (Box 4.9).

Box 4.7 UNEP's 'New Deal for Nature'

In 2019, the United Nations Environment Programme (UNEP) launched the 'New Deal for Nature',[29] spanning 'five transformations'. These include: accounting for the true value of nature; changing the way we produce and consume food; conserving wildlife and wild spaces; restoring the degraded planet; and promoting a better built environment. The New Deal responds to the linkage between decisions affecting biodiversity and economic dimensions. It acknowledges that global capital markets will play a major role in determining whether the UN Sustainable Development Goals can be met.

The 'accounting for the true value of nature' theme of the New Deal for Nature[30] responds to the role of markets and fiscal measures. As amply demonstrated by our history, global markets comprehensively fail to value biodiversity, with under-valuation and over-exploitation of natural ecosystems precipitating a situation in which biodiversity is systematically degraded whilst being taken almost entirely for granted. In particular, the widely applied Gross Domestic Product (GDP) metric totally fails to include the various values of biodiversity or to measure wellbeing and sustainability. It was never designed to do so.

---

[29] UNEP. (2019). *A New Deal for Nature*. United Nations Environment Programme, Nairobi. https://wedocs.unep.org/bitstream/handle/20.500.11822/28333/NewDeal.pdf?sequence=1&isAllowed=y.

[30] UNEP. (2019). *A New Deal for Nature—Account for the True Value of Nature*. United Nations Environment Programme, Nairobi. http://wedocs.unep.org/handle/20.500.11822/28331.

The New Deal for Nature aims to correct this fundamental market failure through economic tools reflecting the true value of nature. These tools include national economic accounting and assessment metrics, and a shift in incentive structures that currently stimulate pollution, ecosystem degradation and over-exploitation of natural resources. Inclusive Wealth Index is one of these tools, developed by UNEP to improve evaluation of progress towards sustainability by accounting not just for manufactured capital but also human capital (the stock of the labour force's skills) and natural capital (natural assets including forests, rivers, land, minerals and oceans) including their contributions to human livelihoods through ecosystem services. The New Deal will also highlight perverse subsidies that currently reward, for example, excessive use of fertilizers in agriculture or overfishing, whilst exploring financial instruments such as tax breaks and payment for ecosystem services that promote pro-conservation behaviours.

Box 4.8 The Regenerative Communities Network

The Regenerative Communities Network[31] is an initiative of the Capital Institute. The Network was founded in 2010 as "a collaborative network working to explore and effect economic transition to a more just, regenerative, and thus sustainable way of living on this earth through the transformation of finance". It aims to promote a global learning network of communities putting into effect a more regenerative approach to development. At its heart is the use of economic systems to contribute to societal good rather than narrowly to make profit, mirroring patterns and principles of natural systems.

'Regenerative communities' in this sense are defined as those pursuing alternative economic pathways that are by their nature diverse, and nuanced to geographical and cultural contexts. The focus of the Network is on the power of economic systems and financial capital to achieve positive social change, emulating natural systems and their inherent renewability rather than the exploitation-based paradigm of neoliberal markets. The Regenerative Communities Network spans members across 6 bioregions, with a focus on community level with the goal to "...accelerate a global just transition from extractive capitalism to regenerative economies one community at a time".

---

[31] http://capitalinstitute.org/regenerative-communities/.

> ## Box 4.9 The Regenerate Costa Rica Hub
>
> At national scale, Costa Rica is engaging with the Regenerative Communities Network through a Regenerate Costa Rica Hub.[32] This approach is guided by the 'Planetary Boundaries' model setting out a set of nine boundaries—climate crisis, ocean acidification, ozone depletion, nitrogen cycle, phosphorus cycle, freshwater use, deforestation and other land use changes, biodiversity loss, particle pollution of the atmosphere, and chemical pollution—within which humanity can theoretically continue to develop and thrive for generations to come.[33] The principal solution posited by the Regenerate Costa Rica Hub is *"Large scale interventions through social empowerment and organization and co-creation, linked to regeneration of functional landscapes"*, the basis of development processes starting with regenerating functional landscapes.
>
> This emphasis on functional landscapes is central, using scientific analysis to determine best land uses to simultaneously optimise the multi-beneficial outcomes of agro-ecological processes that reduce or eliminate chemical fertilisers and pesticides, fix large amounts of carbon to increase organic matter and water retention, and contribute to more efficient production that also allows the recovery of biodiversity and catchment processes. In practice, implementation of regenerative development requires a close participation of farmers, local communities and different experts from diverse disciplines capable of transdisciplinary work, including according high importance to spiritual components. A layered approach looks at systemic links between discrete interests and disciplines, with the longer-term goal of turning Costa Rica into a country that leads on regeneration, demonstrating not only that it is possible but also scalable.

Notwithstanding the dysfunctionality of unconstrained, neoliberal markets, correctly framed market instruments undoubtedly have roles to play towards regenerative change. We have seen this in a variety of market-based instruments in the preceding chapter, for example under 'payment for ecosystem services' (PES) arrangements at scales from the local to the international (as in the case of REDD+). A further example is that of the Climate Investment Fund (Box 4.10), simultaneously addressing climate stability and international equity with biodiversity and a range of linked co-benefits.

---

[32] Universidad para la Cooperación Internacional. (2018). *Regenerate Costa Rica Hub: Recovering the Safe Operating Space for Humanity Through Carbon Capture, Biodiversity and Wellbeing*. Universidad para la Cooperación Internacional.

[33] Stockholm Resilience Centre. (2015). *The Nine Planetary Boundaries*. The Stockholm Resilience Centre, York. https://www.stockholmresilience.org/research/planetary-boundaries.html.

Box 4.10 The Climate Investment Funds (CIF)

The Climate Investment Funds[34] (CIF) was established in 2008, routing funds from donor countries to up-scale climate mitigation and adaptation action in developing and middle-income countries. These public resources are held in trust by the World Bank, disbursed as grants, low-cost loans and risk mitigation instruments to recipient countries through multilateral development banks (MDBs), which add further value by leveraging financing and securing policy support to promote climate-friendly growth. CIF support for transformational change toward low-carbon, climate-resilient development spans four programmes: the Clean Technology Fund (CTF); the Pilot Program for Climate Resilience (PPCR); the Forest Investment Program (FIP); and the Scaling Up Renewable Energy in Low-Income Countries Program (SREP). The CIF Transformational Change Learning Partnership (TCLP) defines transformational change in climate action as strategic changes in targeted markets and other systems, with large-scale, sustainable impacts that shift and/or accelerate the trajectory toward low-carbon and climate-resilient development. These attributes are used in design and evaluation of funded programmes.

An independent review of CIF activities to 2018[35] found that 14 developed nations had cumulatively contributed over $8 billion in support between 2008 and 2018, supporting 300 projects across 72 countries. 'Clean development' projects, including promotion of renewable energy, all make contributions to lightening human impacts on climate systems. However, forest-related programmes including synergy with REDD+ and research into the drivers of forest degradation, as well as promotion of resilience in agriculture, fisheries and natural resource use and management, make the most direct contributions to protection and regeneration of foundational ecosystem resources supporting livelihood security.

Wholesale reform of workings of markets, and the political and public faith in their flawed yet deeply societally ingrained rubrics, is an absolute priority for accelerated revolution if we aspire to a sustainable, conjoined future.

---

[34] https://www.climateinvestmentfunds.org/.

[35] Bird, N., Cao, Y. and Quevedo, A. (2019). *Transformational Change in the Climate Investment Funds: A Synthesis of the Evidence, January 2019*. Overseas Development Institute (ODI), London. https://www.odi.org/publications/11273-transformational-change-climate-investment-funds-synthesis-evidence.

5

A Systemic Decision-Support Framework

So how then do we convert systemic insights into systemic practice when it comes to making decisions at every level from international and national policy-making to a local, community-based project at village level?

In this penultimate section of the book, we convert the learning from global case studies where regenerative socio-ecological systems have been achieved into a simple and easily implemented decision-support framework. STEEP serves as a foundational framework, with each component addressed through one key question and four contextual questions relating to interactions with other elements of the socio-ecological system. These questions are outlined in the following five tables, which are populated by examples from case studies as well as referencing some future challenges. (This framework of key and contextual STEEP questions has been used in a study comparing how perceptions and ensuing water management policy and practical interventions have been shaped in contrasting perceived 'wet' and 'dry' global regions, identifying opportunities to take a more process-based approach.[1])

---

[1] Everard, M. and Quinn, N.W. (forthcoming). Landscapes that hold water: Contrasting perspectives and opportunities from perceived dry (Rajasthan) and wet (England) climatic zones.

© The Author(s) 2020

M. Everard, *Rebuilding the Earth*, https://doi.org/10.1007/978-3-030-33024-8_5

5.1 Tables of Key and Contextual Questions Guiding Decision-Making

These STEEP-based tables are not intended to produce simple answers. It would be misleading to suggest that simple answers exist within the complexity of societal systems and their embedded, legacy norms. Rather, the questions are intended to broker enquiry and dialogue around policy, economic, project or other decisions pertaining to resource use or management. Dialogue itself is important, the systemic framework helping open thinking about who should be in the deliberation process, what may have been overlooked in terms of impacts on supportive ecosystems and societal constituencies, what needs can be integrated in 'systemic solutions', and economic efficiencies achievable through connecting often currently narrowly framed budgets associated with different policy areas.

This systemic decision-support framework can thereby facilitate a change in decision-making culture, informed by systemic context in pursuit of genuinely sustainable, resource-conserving and equitable regenerative socio-ecological systems. It is a framework for innovation, and for co-creativity of decisions that optimise benefits to all. At the very least, it can guide progress from current practice towards more systemically connected outcomes, cognisant of shortfalls and areas for further research, reform or improvement.

Social

Key question:
Have outcomes for all interconnected stakeholders including future generations (for example as expressed through the 17 interdependent UN Sustainable Development Goals) been addressed as key determinants of the success of policies, schemes, technical solutions and resource uses, and have their needs been integrated into decision-making processes?

Framing questions	Examples from successful schemes or future challenges
Technological: What technological solutions, be they nature-based or engineered, serve as 'systemic solutions' optimising benefits for all affected stakeholders, and can inform additional mitigation measures reduce or eliminate negative outcomes if these are inevitable?	• Tarun Bharat Sangh (TBS)-driven regeneration of catchments in the Alwar District of Rajasthan, India, through widespread adoption of geographically and culturally adapted water harvesting, extraction and use solutions are responsive to the expressed needs of people at village scale. • Various landscape management solutions, including consensual cessation of grazing and regeneration of vegetative cover on high-slope erosive soils, have regenerated the Ethiopian Highlands delivering a range of societal benefits.
Environmental: Has the broad diversity of services provided by ecosystems been considered as resources foundational to a broad spectrum of societal benefits?	• Regeneration of water systems at village scale and across catchments in Rajasthan, as well as in the Pakistani Punjab, was a central feature underpinning water, food and socio-economic security and opportunity.
Economic: Has the full spectrum of benefits and disbenefits, their distribution across stakeholder groups as well as net societal value, been considered in the decision-making process?	• Catchment management schemes identified in Upstream Thinking, New York City and SCaMP water supply solutions connected farmers with downstream beneficiaries to seek mutually beneficial outcomes.
Political/governance: Are all relevant stakeholders reflected or engaged in decision-making processes, from problem identification, appraisal and selection of options, to ongoing adaptive management?	• Socio-ecological benefits achieved by catchment regeneration require the needs of different villages to be connected via appropriate governance arrangements. • The needs and priorities of local people were central to policy formulations and decision-making in regeneration and poverty alleviation in China's Loess Plateau.

Technological

Key question:
Does technology choice and operation serve the immediate need (anchor service) as well as the needs of all potentially affected stakeholders (a systemic solution), ideally by avoiding negative impacts or through mitigation, with no automatic presumption in favour of any specific heavy engineering or other approach?

Framing questions	Examples from successful schemes or future challenges
Social: Have all affected stakeholders been considered in perceived problems for which technology solutions are being considered, including at options identification, evaluation and ongoing adaptive management regimes?	• Traditional knowledge and practices are acknowledged as legitimate and locally adapted technical solutions to secure social benefits, as for example the first johad constructed by Tarun Bharat Sangh at Gopalpura that took account of traditional knowledge about how best to address lack of water to tackle the primary cause of poor health, malnutrition and poverty.
Environmental: Have all ecosystem services, including ecosystem resilience across spatial and temporal scales, been considered in technology choice, and is there an ongoing adaptive management plan responding to actual outcomes?	• Integrated constructed wetlands (ICWs) in Ireland, measures implemented in China's Loess Plateau and in the Ethiopian Highlands, and the diverse WHSs found across India are examples of 'systemic solutions', flexibly implemented according to local geography and needs, using natural processes to achieve multiple ecosystem service benefits. • Benefits achieved through Natural Flood Management (NFM) arise in different locations to landscape interventions.
Economic: Has the full spectrum of benefits and disbenefits been considered in technology choice, including any mitigating measures, to take account of net societal value and its distribution across stakeholder groups?	• Ecosystem-sensitive farming at Loddington Farm, Hope Farm and the Kellogg Biological Station demonstrate that 'off the shelf' technologies can transform farming regeneratively and profitably if the value of multi-beneficial outcomes is recognised, and changed farming regimes linked to landscape and soil type at the Knepp Estate also demonstrate profitability. • Reinvestment in tropical dry evergreen forest (TDEF) restoration is a technology adding value additional to turbine installation in Tamil Nadu, India.
Political/governance: Is the decision-making process being undertaken at an appropriate scale to engage all affected stakeholders and consider ramifications for ecosystems in deliberations about technology options, choice and ongoing adaptive management?	• Community-based water and livelihood self-sufficiency drives a lot of the successful schemes in rural India, whereas centrally conceived solutions (such as dam schemes often imposed by asymmetric power relationships) tend to erode system resilience.

Environmental

Key question:
Do decision-making processes take full account of the ecosystem processes and services that underpin continued system resilience and provision of these benefits, forming the basis of 'systemic solutions'?

Framing questions	Examples from successful schemes or future challenges
Social: Is ecosystem use and management sympathetic with natural processes, locally and at scales broader than parochial land ownership?	• Chauka design in Laporiya was an innovation adapted to water interception in natural drainage lines irrespective of land ownership. • The needs of farmers as well as downstream beneficiaries of flood risk management were integrated into management of catchment hydrology under the Stroud Rural SuDS Project.
Technological: Is the technological solution designed and operated in sympathy with natural processes, locally and at scales broader than parochial land ownership?	• Management (NFM) solutions are founded on catchment-scale hydrological processes rather than a narrower approach to 'defending' assets at risk. • Potential engineered and traditional technological hybridisation being explored in Rajasthan's Banas catchment seeks synergy with catchment-scale processes.
Economic: Has the value of all ecosystem services been recognised as a core resource generating multiple benefits?	• Restoration of regionally appropriate TDEF forest in degraded environments of the Coromandel Coast of Tamil Nadu generates multiple carbon sequestration, natural medicine, educational, hydrological, biodiversity and other linked ecosystem service values. • The examples of community-based water management solutions in Rajasthan exemplify local-scale beneficial solutions (johadi, chauka, taanka, etc.) attuned to highly localised geographical situations. • At landscape scale, restored ecosystem functioning including water and soil retention achieved through reforestation and re-greening underpins linked economic, environmental and social regeneration in the Loess Plateau of China. • Were recommendations for coastal wetland restoration to be adopted to protect New Orleans from flooding, a diversity of ecological, fishery, amenity and other benefits would arise.
Political/governance: Does the decision-making process take full account of long-term ecosystem integrity and processes underpinning sustainable outcomes?	• Identification and designation of marine protected areas (MPAs) is based on recognition of the need to safeguard marine species and processes of value for ecosystem integrity and sustainable production of societal benefits. • A significant enabler, or alternatively obstacle, to NFM is appropriate legislation, policy instruments and supporting budgets recognising ecosystem service co-benefits.

Economic

Key question:
Are financial incentives and taxes and their economic outcomes informed by benefits and disbenefits across the full spectrum of ecosystem services, and not merely narrowly focused on short-term returns to a subset of often privileged stakeholders?

Framing questions	Examples from successful schemes or future challenges
Social: Have the needs and values of all affected stakeholders been reflected in value judgements, including both benefits and disbenefits?	• Water system regeneration in villages in Alwar and Laporiya (Rajasthan) were driven by the societal benefits of water, food, biodiversity and socio-economic security.
Technological: Are all multiple values, positively as well as negatively, fully accounted for in the proposals for, appraisal and implementation, and ongoing adaptive management, of technological solutions?	• A 'systemic solutions' approach seeking to optimise multi-beneficial outcomes from ecosystem functions is reflected in Upstream Thinking, SCaMP and New York City water supply, Natural Flood Management schemes, and poverty alleviation in solutions in the Ethiopian Highlands.
Environmental: Is net value to society framed across the whole range of ecosystem services and their associated beneficiaries, rather than narrow benefits to a few stakeholders, or a presumption founded on the fundamentally flawed notion of the 'trickle-down effect' (where benefits to the few are assumed to 'trickle down' to all in society but rarely do)?	• Farm profitability remains a priority at the Knepp Estate in the UK, and economic viability is part of the assessment of practices at the Kellogg Biological Station in the US, demonstrating that ecological recovery need not be in conflict with economic returns. • Stabilisation of soils and water through re-greening of the Loess Plateau and Ethiopian Highlands underpins environmental, economic and social benefits. • Successes of Working for Water are rooted in the recovery of native ecosystems generating secure flows of water of substantial economic benefit. • Reanimation of linked ecological and social elements of catchments through ICWs in Ireland and the Upstream Thinking, SCaMP and New York City water supply schemes were planned to deliver multiple benefits beyond their driving 'anchor services'. • Rehabilitation of coastal wetlands would be cost-effective for tackling the vulnerability of New Orleans, also yielding diverse ecosystem service benefits.
Political/governance: Is the decision-making process cognisant of all values, in terms of benefits and disbenefits, accruing from options for resource use or management?	• The Upstream Thinking, New York City and SCaMP water supply solutions balanced tangible benefits to city water supply with livelihood safeguards and funding for land owners and users in source catchments. • Net societal value is part of the innovation and cost-efficient 'reverse auction' arrangements put in place to reduce nutrient loads to Chesapeake Bay.

Political

Key question:
Is legislation and political debate adequately informed about human and ecosystem interdependencies, robust, and sufficiently flexible to allow decision-making to occur through the most appropriate level of formal and/or informal governance, and is there coherence between these political strata to integrate community needs and knowledge with ecosystem processes and resilience?

Framing questions	Examples from successful schemes or future challenges
Social: Has decision-making been delegated to the most appropriate level to account for local geographical and cultural contexts and traditional knowledge, and the roles of people as key resource owners, actors and communicators?	• Village-scale governance through *Gram Sabha* is essential for effective design, operation and continued ownership of groundwater recharge schemes in Alwar, and have informed Government of Rajasthan policies such as *Mukhya Mantri Jal Swavlamban Abhiyan* (MJSA) increasing coherence between levels. • National schemes have been co-designed by local needs and contexts, and delivered by local actors, in China's Loess Plateau and the Ethiopian Highlands projects as well as the aspirations of Natural Flood Management.
Technological: Is technology choice informed by inclusion of all types of knowledge and needs?	• Solutions addressing driving priorities as 'anchor services', around which outcomes for all ecosystem services and hence net societal value are optimised, include carbon sequestration services in forest restoration in Tamil Nadu and cost-efficient water resource protection under Upstream Thinking, SCaMP and New York City's water supply schemes.
Environmental: Have impacts on the processes and services of ecosystems been considered in decision-making, which may entail nested governance arrangements to avert fragmented management?	• Tarun Bharat Sangh's promoted catchment-based *Pad Yatra* ('Water Parliament') as a forum to promote basin-wide water sharing, dispute resolution, water body restoration, soil fertility, reforestation and associated livelihood enhancement between *Gram Sabha* (village councils) in the Arvari and other drainage basins in Alwar District. • Environmental successes achieved through widespread implementation of Integrated Constructed Wetlands (ICWs) in Waterford have informed national policy and become incorporated in Irish Government guidance.
Economic: Has the full spectrum of benefits and disbenefits been integrated in deliberative processes, incentives and taxes, with opportunity to challenge policies that block optimally sustainable outcomes?	• Working for Water in South Africa linked up a range of social, economic and environmental goals and budgets across government departments to create a coherent, multi-beneficial and widely supported programme. • Tarun Bharat Sangh in India and Zephaniah Phiri in Zimbabwe experienced government opposition but persisted to introduce and spread effective water resource regeneration methods, and communities in Alwar District denied access to a contractor awarded a fishery contract by state government in their resource water bodies.

5.2 Systemic Interconnections as Part of an Ongoing Societal Transition

As with any system, all aspects of the socio-ecological system have to be addressed in an integrated way. 'Cherry picking' elements of the system in isolation, for example a policy environment predicated on maximising economic productivity in the short term without regard to environmental and social consequences, is likely to favour technology choices undermining overall system integrity, equitable benefit-sharing and long-term sustainability. Conversely, regenerative outcomes are possible when all facets of the system—in this case articulated through the STEEP framework—inform policy decisions, choices and technology deployment that promote long-term sustainability, equity and economic viability. When this 'virtuous circle' is achieved, many co-benefits can result.

So how does transformation of societal norms occur? A study has described how the 'ethical envelope' of society's norms and expectations has expanded considerably throughout history, including embracing the rights of more sectors of humanity as well as emerging pro-environmental values, progressively institutionalised as consensual agreements in the form of regulatory, fiscal, market, social protocol and other mechanisms.[2] The study highlighted the importance of a 'ripple effect', wherein these issues become progressively recognised not as narrowly 'environmental' but in terms of their wider health, social, economic and other 'non-environmental' implications. This 'ripple out' into broader societal consequences subsequently directs and accelerates the pace at which environmental concerns begin to shape mainstream societal norms and structures. Many examples arise from the knowledge base in this book, such as the influence of once-contested community-scale water solutions (such as ICWs, johadi and Phiri pits) that are now integrated into state policies.

This systemic consideration of all factors is essential not merely to drive a transition in regenerative thinking, but to maintain momentum

---

[2] Everard, M., Kenter, J.O. and Reed, M. (2016). The ripple effect: Institutionalising pro-environmental values to shift societal norms and behaviours. *Ecosystem Services*, 21, pp. 230–240.

after decision-making and implementation. Successes may conflict with fixed assumptions, vested interests and legacy regulations but, if their breadth of socio-ecological outcomes are increasingly recognised, can provide a learning resource for the progressive reform of the policy environment (see Box 5.1).

Box 5.1 Examples of Policy Learning from Successful Regenerative Schemes

- Sustainable farming techniques being implemented at Loddington Farm, Hope Farm and the Kellogg Biological Station are monitored across a range of economic, nutrient, soil retention and ecological parameters, most of them unaccounted for in the mainstream of intensive farming practice, to continuously improve methods and lessons for wider uptake. These multi-beneficial effects, for example including lower carbon intensity, can inform policy decisions towards more sustainable land uses
- The first johad constructed by Tarun Bharat Sangh (TBS) at Gopalpura was a high-risk undertaking, but positive outcomes generated local interest. TBS, as well as Zephania Phiri in Zimbabwe, both resisted state authority, which took out civil sanctions against them, to assert local control of water management and regenerative outcomes. In both cases, as well as in the positive outcomes driven by NGOs such as WaterHarvest, the societal benefits of these successes have subsequently been recognised as coherent with state goals of rural water sufficiency (particularly the *Mukhya Mantri Jal Swavlamban Abhiyan* programme in Rajasthan) and national policies.
- Evident benefits for river habitat diversification and functioning generated by American beaver reintroductions are now being applied as part of conservation measures for endangered Chinook and Steelhead salmon in the Pacific Northwest of the USA, promoting fishery management and angling tourism goals, crossing policy area boundaries.

An adaptive management approach is also essential to ensure that practical outcomes, particularly social and environmental and associated distributional economic outcomes, inform revision of strategy, to ensure that technological and policy interventions remain systemically connected. As the knowledge base in Chap. 3 of this book demonstrates, learning outcomes from regenerative schemes also provide a valuable resource informing future projects and policy reform as part of a wider societal culture change.

5.3   Relevance to Decision-Making Across Major Policy Areas

Application of this decision-support framework and the systems thinking behind it does not apply narrowly to 'environmental' issues that, as we have observed, are anyway spuriously defined. Rather, as ecosystems and their myriad services and interdependences underpin all dimensions of human wellbeing, it is germane to all spheres of human activity. This includes all of the major policy areas into which governance tends to be classified across the world and, importantly, the often neglected connections between them.

My 2016 book *The Ecosystems Revolution*[3] provided examples of the central and often underappreciated roles played by ecosystems in all policy areas, and considered how further integration of ecosystems thinking could deliver greater and more optimal outcomes. Departmental divisions common to governments globally included: Treasury (perhaps the most influential of all government departments); Business and Trade (some of the implications of which we have already explored in terms of leadership for natural resource stewardship); Energy; Urban Planning (we have also considered the benefits of 'greener' cities); Transport; Agriculture and Food (transitions and aspirations for farmed land have been addressed previously); Health and Wellbeing; Natural Environment (covered in many places throughout this book); International Development; Defence; Foreign Policy; and Research and Education. As the relevance of ecosystems thinking may appear less intuitive for some departmental interests than others, Box 5.2 summarises examples for Defence, Foreign Policy also including International Development, and Transport.

---

[3] Everard, M. (2016). *The Ecosystems Revolution: Co-creating a Symbiotic Future*. Palgrave PIVOT Series. Springer, Switzerland.

Box 5.2 Examples of Relevance of Ecosystems Thinking to Selected Policy Areas

- **Defence**: Whether dressed as ideology, race, faith or other, conflict usually has competition for a limited resource at its core. Oil has been a commonly contested resource in our more recent industrial past. The 2009 prediction by then World Bank Vice President Ismail Serageldin that *"Many of the wars of the 20th century were about oil, but wars of the 21st century will be over water unless we change the way we manage water"*[4] is today still often repeated. The reality is that there has been a long history of water-related conflicts. Examples include the 1967 Six Days War in the Middle East, drainage of the Mesopotamian wetlands as a potent weapon of war waged by Saddam Hussein's forces against Iraq's Marsh Arabs, and underpinning simmering tensions related to sharing of major transboundary rivers between India, Pakistan and China.[5] The 2006 UN report *Water Wars to Bridges of Cooperation—Exploring the Peace-Building Potential of a Shared Resource*[6] catalogues the role of water resource-sharing and co-management in international security across the globe. Many more examples of water-based cooperation with associated peace-keeping and peace-making are reviewed in my 2013 book *The Hydropolitics of Dams*.[7] Water and other vital natural resources cannot be dissociated from the securitisation agenda, a reality not lost on the military. The 2018 (sixth) edition of the UK Ministry of Defence's forward look document, *Global Strategic Trends: The Future Starts Today*,[8] includes 228 instances of the word 'water', 9 of 'soil' and 101 of 'food', highlighting substantial and growing awareness of foundational ecosystems in the global security agenda.
- **Foreign Policy** also including **International Development**: There is growing awareness that the pressing sustainability challenges facing the world are truly playing out at large scale. Actions and influences of

---

[4] Serageldin, I. (2009). Water: Conflicts set to arise within as well as between states. *Nature*, 459, p. 163.

[5] The Economist. (2011). Unquenchable thirst. *The Economist*, 19 November 2011. http://www.economist.com/node/21538687, accessed 15 October 2015.

[6] United Nations Department of Public Information. (2006). *Ten Stories the World Should Hear More About: From Water Wars to Bridges of Cooperation—Exploring the Peace-Building Potential of a Shared Resource.* http://www.un.org/Pubs/chronicle/2006/issue2/0206p54.htm#Water.

[7] Everard, M. (2013). *The Hydropolitics of Dams: Engineering or Ecosystems?* Zed Books, London.

[8] Ministry of Defence. (2018). *Global Strategic Trends: The Future Starts Today*, 6th ed. Ministry of Defence, London. https://assets.publishing.service.gov.uk/government/uploads/system/uploads/attachment_data/file/760099/20181121-GST_The_Future_Starts_Today.pdf, accessed 16 January 2019.

richer countries have compromised people in poorer nations meeting their needs. More positively, international cooperation has mobilised nations around global threats such as climate change, species loss and degradation of marine resources, as well as implications for international supply chains.

- Transport:

  - Managers of transport infrastructure have traditionally failed to make connections with their ecosystem dependencies and, downstream, their broader consequences for the environment (beyond growing concerns about climate-active gas emissions). Issues such as the flooding of road and rail infrastructure have instead generally been perceived and managed as a local problem, to be dealt with by installing bigger drainage pipes and/or pumps.[9] Yet there is now a slowly dawning awareness that transport routes have generally been aligned for convenience, for example railway lines built as straight as possible across drainage basins and in tunnels that cut through aquifers. Hard lines could be beneficially 'softened' to account for natural processes following landscape topography, potentially averting the flooding of road and rail embankments, cuttings and tunnels, and their contributions to subsequent downstream flood risk.

  - Historic bridge design tended to constrain river flows and the mobility of river channels within their floodplains. This makes bridge supports vulnerable to erosion, contributes to flooding particularly where narrow bridge spans become obstructed by debris, and disrupts geomorphological (habitat-forming) processes and the free passage of fish and other wildlife. Improved bridge design better informed by natural processes provides space for flood flows, mobility of river channels that tend to meander within their floodplains, and greater connectivity across the river corridor allowing continuity of natural functions and ecosystem services (Fig. 5.1).

  - Also, cycling, walking and other low-impact transport routes can be designed as 'systemic solutions', co-delivering many linked, cross-policy ecosystem benefits such as visual and noise buffering, wildlife corridors, recreational areas and natural urban drainage systems.

[9] Quinn, N.W., McEwen, L.J., Parkhurst, G., Parkin, J., Everard, M., Horswell, M., McInnes, R.J., Newman, R. and Williams, D. (2015). *Co-creating Railway Flood Resilience: Applying the Science of Blue-Green-Grey Infrastructure*. NERC, 'Environmental Risks to Infrastructure Innovation' programme, University of the West of England.

Fig. 5.1   This bridge over the tidal River Suir upstream of the city of Waterford, Ireland, provides space for the mobility of the river within its floodplain and connectivity between the river channel and floodplain, safeguarding wildlife and ecosystem services. (Image © Dr Mark Everard)

Perhaps most important of all is recognition of the systemic connections between all policy areas. Opportunities arise from focusing on underpinning ecosystem processes as a basis for more sustainable and multi-beneficial 'systemic solutions', optimising outcomes across ecosystem services, policy areas and public budgets when recognised as part of the same integrated system. At worst, closer integration may provide economic efficiencies whilst also revealing and potentially helping avert unintended negative cross-policy outcomes. At best, it provides a basis for more profound societal transformation influencing management strategies, decisions and practices integrating supporting ecosystem processes into daily decisions.

Wholesale change is undoubtedly necessary if the collective lifestyles of humanity are to move onto a more regenerative pathway, valuing and restoring the supportive capacities of currently much degraded global ecosystems as the primary resources underpinning human security and opportunity.

6

Epilogue: Rebuilding the Earth—Yes, We Can!

This book is about culture change, concerning both the necessity but also the means to put nature and people back into the heart of societal thinking, policy and action.

It is about sustainable development in its deepest green sense, pragmatically embedded in social, technological, governance and economic contexts as a spur to innovation. The repeating meme is systemic interconnection, a biophysical reality albeit requiring the transformation of today's fragmented societal norms if we are to navigate our way towards a sustainable path.

This book is also about optimism. Yes, we can! In fact, we are already making regenerative steps in fragmented ways across the world. Whilst it is easy to feel overwhelmed by the yawning reality of booming cultural and environmental threats, this can only stultify action and is therefore a luxury we can ill afford. Instead, this book focuses not merely on the fact that we can make regenerative change, but that we have already done as witnessed through many case studies globally. These, in turn, can teach us much about how to rethink our decisions at international, national, state and right down to highly localised scales—urban, rural, agricultural, industrial—and particularly how to achieve coherence between these interconnected facets and scales (Fig. 6.1).

© The Author(s) 2020
M. Everard, *Rebuilding the Earth*, https://doi.org/10.1007/978-3-030-33024-8_6

Fig. 6.1 The over-riding challenge facing a booming human population subsisting on diminishing ecosystems is to find a sustainable accommodation founded on rebuilding natural systems as the most vital resource underpinning human health, wealth and life fulfilment. (Image © Dr Mark Everard)

So we return to the metaphor of 'Rebuilding the Earth': the natural world and its processes that constitute resources essential for a future of security and opportunity. There is unquestionably an urgent need to safeguard our inheritance of a natural world that is still functional, if substantially degraded by recent human development, then to set about the task of facilitating its regeneration through progressive shifts in attitude and practice across society.

We need to reconstruct the metaphorical 'ark' of natural species, processes and services for inherent purposes, but also for self-interest as this is the only vessel we have, upon which we are totally dependent for our journey into a future we still have the option of wilfully enriching.

This book is dedicated to you, the reader, who has more power and influence than you may yourself believe in promoting systemic and sustainable change in the societal sectors and worlds with which you interact. We all have the power and duty to drive forward regeneration of the 'ark' of planetary ecosystems and the human prospects that it bears. We all have capacities to rebuild the Earth for a sustainable future.

Glossary

Anchor service An ecosystem service that is the desired focus of ecosystem use or management, that can serve as an 'anchor' around which consequences for other interlinked ecosystem services are assessed and, where possible, optimised.

Anicut A low dam across gently sloping land built to retain water during monsoon flows.

Biotic homogenisation Introduction of non-native species and the extinction of local biodiversity, increasing the genetic, taxonomic or functional similarity of ecosystems in different locations.

Chana Chick pea (Hindi).

Chauka 'Rectangle' (Hindi) pits dug in Laporiya to intercept monsoon run-off.

Clean production A preventive initiative intended to minimize waste and emissions and maximize product output.

Corporate Social Responsibility (CSR) A self-regulating and generally transparently reported business model intending to help a company to be socially accountable.

Dahl Lentil (Hindi).

Degenerative landscapes Landscapes and waterscapes in which degrading ecosystem integrity, functioning and services also inevitably degrade linked socio-economic security and opportunity.

Ecosystem services The multiple, diverse but often overlooked benefits that ecosystems provide to people.

© The Author(s) 2020
M. Everard, *Rebuilding the Earth*, https://doi.org/10.1007/978-3-030-33024-8

Externalities Unintended negative outcomes.

Flood-retreating cropping A common production method in dry regions with seasonally variable rainfall, exploiting stored soil moisture in the margins of water bodies for cropping as water levels recede.

Gram Sabha Village council (Hindi).

Gram Vikas Navyuvak Mandal, Laporiya (GVNML) An NGO (translating as 'village growth youth board, Laporiya') based in the village of Laporiya (in the Jaipur District of Rajasthan state, India) working on water and ecosystem management of community security.

Heat island The higher temperature encountered in built-up areas compared to nearby rural areas.

Integrated constructed wetland (ICW) Man-made wetland systems comprising cascades of shallow wetland cells optimised to produce a linked range of ecosystem services including, for example, water purification, amenity, support for biodiversity and aesthetic improvement.

Jal Bhagirathi Foundation Indian NGO with interests in promotion of water self-sufficiency, primarily working with rural communities.

Jalyukt Shivar Abhiyan (JSA) Water self-sufficiency programme led by the Government of Maharashtra.

Jayad Third (summer) crop in Indian drylands.

Johad (plural 'johadi') A generally semi-circular dam built across a drainage line in the landscape to intercept monsoon run-off.

Khariff Wet, post-monsoon crop in Indian drylands.

Land sharing Combining biodiversity conservation with food production, generally at lower intensity and yield, on the same parcels of land.

Land sparing Segregation of areas protected for wildlife from those devoted to more intensive agricultural production and other forms of development.

Managed realignment Also sometimes known as 'managed retreat', this is the process of deliberately breaching artificial seawalls to allow the reinstatement of tidal inundation over areas of formerly 'reclaimed' low-lying land often converted in past decades or centuries for agricultural purposes.

Naadi A low bund surrounding fields on land with a low slope (Hindi).

Nexus A linked set of interacting parameters, often in sustainable development discourse highlighting the integrally interlinked nexus of food, water and energy that may limit human development.

Nullah Drainage lines generally in sloping land (Hindi).

Out-scaling Promotion of wider geographical pervasion of successful initiatives.

Payment for ecosystem services (PES) A market-based instrument in which the beneficiaries of ecosystem services make payments to ecosystem stewards influencing the provision of those services.

Rabi Second, dry season crop in Indian drylands.

Regulatory lag Time lag entailed in revision of the regulatory environment to reflect evolving understanding and priorities.

Ripple effect Issues initially perceived narrowly as being 'environmental' or 'social' tend only to become better and more widely appreciated, and consequently institutionalised in legislation or other societal protocols, when their implications 'ripple out' across other societal interests (e.g. health, economy, risk, etc.).

Sagar Surface water body (Hindi).

Socio-ecological systems (SES) Tightly linked ecosystems and the socio-economic activities and prospects of people dependent upon and also influencing them.

Sodic Containing a high concentration of sodium.

Sustainable Catchment Management Programme (SCaMP) Water quality protection initiative in the north-west of England based on landscape management.

Sustainable intensification A process or system where agricultural yields are increased without adverse environmental impact and without the conversion of additional non-agricultural land.[1]

System A complex whole (a cell, a universe, an atom, a watch, a corporation, etc.) comprising interacting or interdependent component parts, each surrounded and influenced by its environment and other elements of the system.

Systemic solutions: *"…low-input technologies using natural processes to optimise benefits across the spectrum of ecosystem services and their beneficiaries"*.[2]

Taanka A water-harvesting structure (WHS) adapted to water capture in flat, arid lands.

Tank Monsoon water interception and storage system.

Tarun Bharat Sangh (TBS) A Gandhian-based NGO based in Alwar District (Rajasthan state, India) working on water and ecosystem management of community security.

---

[1] Pretty, J. and Bharucha, Z.P. (2014). Sustainable intensification in agricultural systems. *Annals of Botany*, 114(8), pp. 1571–1596. https://doi.org/10.1093/aob/mcu205.

[2] Everard, M. and McInnes, R.J. (2013). Systemic solutions for multi-benefit water and environmental management. *The Science of the Total Environment*, 461(62), pp. 170–179.

Tipping point The point at which a series of small changes or incidents becomes significant enough to cause a larger, more important change.

'Trickle-down effect' Where benefits to the (generally privileged) few are assumed (generally falsely) to 'trickle down' to all in society.

Up-scaling Replication of successful schemes at larger scale to increase overall benefits.

Upstream Thinking Water quality protection initiative in south-west England based on landscape management.

WaterHarvest The rebranded name, adopted in 2017, for the community-facing NGO Wells for India.

Wells for India A community-facing NGO established initially as a UK-based charity in 1987 with water management, sanitation and self-sufficiency goals, with an operational base in Udaipur (Rajasthan, India).

Index[1]

---

[1] Note: Page numbers followed by 'n' refer to notes.

© The Author(s) 2020

M. Everard, *Rebuilding the Earth*, https://doi.org/10.1007/978-3-030-33024-8